General Chemistry

Laboratory Manual and Notebook

Using Biochemical Tools

Fourth Edition

Philip N. Borer

Syracuse University

Kendall Hunt
publishing company

Cover image © Shutterstock, Inc.

Kendall Hunt
publishing company

www.kendallhunt.com
Send all inquiries to:
4050 Westmark Drive
Dubuque, IA 52004-1840

Copyright © 2005, 2006, 2010, 2012 by Philip N. Borer

ISBN 978-1-4652-0562-9

Printed in the United States of America
10 9 8 7 6 5 4 3 2 1

Contents

Note to the Student

This manual contains experiments that will be conducted in the General Chemistry laboratory. Please see the schedule in the syllabus for more details.

Recording Data. Use the **Notes** pages after the written part of each lab module to record your observations.

Reports. After lab, do the calculations and answer the questions on **Report** pages that will be distributed in the lab and are available at the course Blackboard web site. To a large extent, your grade will be determined by the quality of your reports.

Always check the Blackboard site as your first resource for any question. It will list which experiment will be conducted in a given week and communicate any last-minute changes. Your assigned TA also will give you assistance.

The University has made a very substantial investment in equipping the General Chemistry laboratory with tools used in modern biochemical research laboratories. We hope you enjoy this unique privilege.

Most student participants in this course have already had high school chemistry. A few have not, so some of the early experiments review topics that are important to make the work accessible to everyone.

The laboratory staff will do its best to resolve problems as quickly as possible. Please let your TA know if there is something unclear in the instructions, so next year's class can benefit from your experience.

Laboratory 1

For Your Safety

Adopt the attitude that **ACCIDENTS WILL NOT HAPPEN!** You are NOT going to do anything, or let anything happen to you, that will require medical attention.

Work in a chemical laboratory has potential hazards. It is your responsibility to learn about them and take appropriate precautions. Otherwise, you risk serious injury or death to yourself and others. Common sense and knowing where to get help are your most important allies. The list of items that follow aims to sharpen your safety awareness.

1. Once your work area has been assigned, write down the nearest location of the safety items that follow. Quick, common-sense action can stop a serious accident. Your instructor will point out the locations and explain the use of:

    15mins, if required

    Eyewash stations

    Emergency showers   *Directly behind me. Near the door.*

    Fire extinguishers

    Fire blankets

    Fire alarms

    Emergency phone

    Emergency exits

    First aid station

    MSDS sheets   *Red binder*

BEFORE STARTING AN EXPERIMENT

2. You will be sent home to change if you show up in shorts, short skirts, bare feet, sandals, or open shoes. Don't wear loose clothing that may brush against a container and cause a spill. It is suggested that you wear old jeans or sweatpants and a snug-fitting shirt or blouse, rather than expensive clothing.

3. Turn off cell phones before entering lab. Wash your hands prior to using a cell phone outside the lab.

4. Read the experiment before coming to lab. Note questions you may have, and ask the instructor if they are not clarified in the pre-lab lecture.

5. Pay close attention to the directions your instructor gives you. The instructor may change some aspects of the experiment—listen to his/her pre-lab announcements.

6. The written description of the experiment has specific safety precautions. Pay careful attention to all hazard warnings, and ask questions if you are confused about any aspect. The instructor will tell you the location of the Material Safety Data Sheets (MSDS) for chemicals in use in the lab. You have the right to examine these sheets if you wish. MSDS information includes physical properties; health hazards; reactivity, fire, and explosion data; procedures to use in case of spillage; disposal information; and any special precautions.

7. If you have serious allergies to chemicals, you should probably not take this or any other chemistry laboratory course. Inform your instructor if you have mild allergies or another medical condition that may require you to limit your exposure to small quantities of chemicals.

8. Do not use the stools in the lab. They present a hazard in case we need to evacuate in a hurry. Do not sit on the benches or put personal items on the bench tops. A person in another lab section could have contaminated the bench with chemicals.

9. Do not eat or drink in the lab.

WORKING IN THE LAB

10. _WEAR GOGGLES_ continuously when the instructor says to begin wearing them. The instructor may eject you and give you a zero for the lab report if you do not wear them. Do not wear contact lenses—chemical vapors can irritate your eyes, and contacts may make it hard to flush your eyes in case of a spill.

11. Avoid touching hot objects. Remember that heated objects may remain too hot to hold for 10–30 minutes—this includes heating blocks, hotplates, etc.

12. Avoid breathing fumes. If you need to detect the presence of an odor, hold the tube about a foot from your face and use your hand to gently waft some of the fumes toward your nose. Never inhale fumes from a test tube directly beneath your nose.

13. Do not use damaged items. Contact the instructor for a replacement.

14. Never work alone in the laboratory. An instructor must be present at all times.

15. Work only on the assigned experiment. Ask the instructor before changing anything in the text or verbal instructions.

16. "Drive Defensively." Watch what your neighbors are doing. Do not be afraid to tell them to be more careful. If they continue dangerous activities, notify the instructor. Do not fool around. Be serious, not somber—chemistry can be interesting and fun.

WASTE DISPOSAL AND CLEANUP

17. Dispose of chemical wastes in the containers provided for specific materials. Follow the instructor's directions.

18. Put **_BROKEN GLASS_** in the proper container in the front of the lab—**_NEVER_** put glass in the common trash.

19. Keep your lab space clean. Use a sponge and water to clean your bench if it is dirty. The instructor will refuse to sign your Notes if you need to clean up your work area.

20. Wash your hands with soap and water before eating, getting a drink, using the bathroom, or leaving for the day. Do not touch your face without washing your hands first.

IN CASE OF AN ACCIDENT

21. Report any injury, no matter how minor. Stay calm. Always get help! This is no time to worry about your grade in lab.

22. If a **_CHEMICAL SPLASHES IN YOUR EYES,_** immediately call for assistance and let someone help you to the eyewash station. Keep your goggles on until you reach the eyewash station or you could splash from your goggles to your eyes and make matters worse. Quickly drench your face and goggles, then flush your eyes thoroughly—about ten or twenty minutes. Force your eyes open and rotate your eyeballs to be sure water reaches everywhere.

23. If a **_CHEMICAL SPLASHES ON YOUR SKIN_**, immediately flush the area with cold water. Have someone notify your instructor. If the spill affects a large area of your skin, go quickly to the safety shower, remove all clothing that may be contaminated and flood the affected area with cold water. Have someone notify your instructor. In some cases, he/she will recommend washing with a mild detergent followed by extensive rinsing with cold water.

24. Notify your instructor immediately if you or anyone near you ingests a chemical.

25. For **_MINOR CUTS_**, wash the area with water and report to the instructor for a bandage.

26. If there is **_SERIOUS BLEEDING_**, apply direct pressure with a clean, preferably sterile, dressing. Have someone immediately call campus emergency. Notify your instructor.

27. For a **_MINOR BURN_**, let cool water run over the burned area. This applies to burns from chemicals, hot objects, or flames. Call for your instructor to see what to do next.

28. **_EXTINGUISHING A FIRE_** is done by directing the fire extinguisher spray at the base of the flames. When possible, use short bursts to avoid depleting the extinguisher. Call for someone to shut off or remove the fuel that is supplying the fire. Notify the instructor and call campus emergency. Everyone else should evacuate the area immediately.

29. **_STOP-DROP-ROLL_** if your **_CLOTHING IS ON FIRE_**. This means **_STOP_** what you are doing, **_DROP_** to the floor, and **_ROLL_** over and over to extinguish the flames. Roll away from the source of the flames. Someone will bring a fire blanket quickly. **_NEVER_** run to the fire blanket or shower. That increases the oxygen supply to make the flames burn hotter and makes you breathe faster—that can make you inhale toxic fumes and hot gases causing severe respiratory damage. Restrain panic as much as possible—people will come to your aid. Your instructor may move you to the safety shower.

30. If you ***SEE SOMEONE'S CLOTHING ON FIRE***, immediately grab a fire safety blanket and cover the person to extinguish the flames. Remove the fire blanket when the flames have gone out. Call for someone to bring the instructor and call campus emergency immediately. Call for someone to extinguish the fire.

31. If you ***SEE SOMEONE UNCONSCIOUS***, immediately call the instructor. Examine the area for potential electric shock hazard or other dangers. Call campus emergency immediately.

32. If there is a ***CHEMICAL SPILL*** on the bench, floor, or a reagent shelf, notify your neighbors and your instructor. If the spill can explode or burn, call for everyone to evacuate. Let your instructor supervise the cleanup, and/or bring an extinguisher.

Read the ***Conditions for Safe Participation in the Chemistry Laboratory*** on the next page, sign it, then take the safety quiz. You will not be allowed to work in the lab until you sign the conditions and you get 100% of the quiz correct.

Conditions for Safe Participation in the Chemistry Laboratory

I agree to the following conditions regarding my participation in the chemistry laboratory:

1. I agree to **wear approved safety goggles** at all times required by my instructor.
2. I will **turn off my cell phone** prior to entering the lab and refrain from using it throughout the lab session.
3. I am responsible for knowing the **location and operation of all safety equipment** in the laboratory, including emergency exits, eyewash stations, safety showers, fire extinguishers, fire blankets, fire alarms, and emergency telephones.
4. I will **wear appropriate clothing** in the lab. This means I will not wear shorts, short skirts, open-toed shoes, bare feet, or loose clothing. I understand that the **instructor may send me home** to change if I am not dressed appropriately.
5. I will **read the experiment prior to arriving** in the laboratory, be prepared to take a quiz related to the experiment, and will ask questions if something is not clear.
6. I am responsible for storing **personal items** in an appropriate place—I will not leave them where they might distract myself or others who are working in the lab. This includes backpacks, coats and clothing, extra books, briefcases, laptop computers, cell phones, etc.
7. I will **never eat or drink** in the laboratory.
8. I will **never work alone** in the laboratory.
9. I will **work only on the assigned experiment.**
10. I will **use a fume hood** when instructed to do so.
11. I will **check the labels** on chemical items prior to dispensing them into my own containers. I will inform my instructor if it is likely that any stock solutions have been contaminated with materials that are not on the labels.
12. I will inform my instructor if I discover a **spill of a chemical** on a bench or the floor.
13. I will **avoid touching hot objects and hazardous chemicals.**
14. I will **properly dispose of waste materials** according to the instructor's directions.
15. I will **immediately report any dangerous situations** or injuries to my instructor.
16. I will **clean up my workspace** with a sponge and water prior to leaving each lab session.
17. I will **wash my hands** thoroughly prior to leaving the laboratory.

Name _Brandon Breazeale_

Section Number _M014_

Instructor's Name _Cara Rufo_

I declare that I have read this safety agreement and agree to each of the conditions named herein. I understand their importance for my own safety and for the safety of others in the laboratory.

Brandon Breazeale   _11:00 AM_

student's printed name   meeting time

[signature]   _9/13_

student's signature   date

_____   _____

instructor's signature   date

Laboratory 2

Volume and Mass Measurements

We all want more effective medicines, safer food and water, and longer lasting materials in airplanes, cars, etc. These are desirable qualities—we usually think about such issues in a *qualitative* manner—but often the key to progress in understanding a system is to think about its *quantitative* description. Chemistry and many other disciplines are experimental sciences. A properly designed experiment will tell us *what* and *how much*.

Understanding the source of errors in measurements is very important in experimental science. New discoveries often occur when a scientist first observes small deviations from expected values. *Random* errors tend to average out over many measurements. Sometimes, this kind of error is called *noise*. Often it is possible to reduce the noise and improve the precision of repetitive determinations by changing experimental conditions. Deviations from the accurate (true) value of a measurement also occur due to *systematic* errors. A scientist should do everything possible to eliminate systematic errors, which skew a measurement in one direction from the true value.

The module on **Dimensional Analysis** will describe moles and numbers of molecules and how these are related to masses and volumes. In this module you will learn to use a microliter pipet to measure the volumes of aqueous solutions. At the same time, you will learn about what it means to measure masses and volumes accurately. Please review the sections in your textbook that describe precision, accuracy, and significant figures.

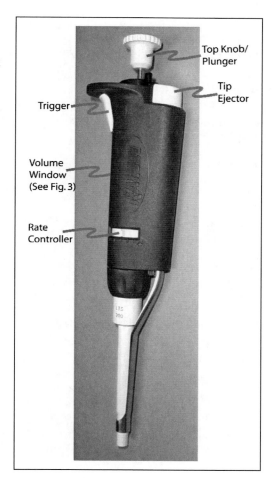

FIGURE 2-1
Side view of a P-200 Latch-Mode Pipet.

You and your partner will be using a modern pipet that is capable of measuring microliter volumes accurately. These pipets are widely used in biochemical and chemical research laboratories. You will also be using delicate balances for weighing samples.

BE CAREFUL!

These are precision instruments that can be damaged by thoughtless users. Be sure to understand the directions printed here and those given by your instructor. We hope you enjoy the experience of working with this state-of-the-art equipment. Please act in a professional manner and return the equipment in good working order.

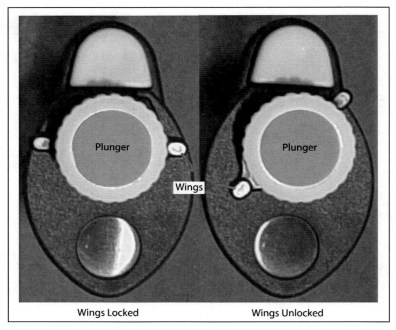

FIGURE 2-2
Top view of the pipet.

Recording Data. There are pages at the end of the lab module to record your observations.

1. **Latch-Mode MicroLiter Pipet, 0–200 µL**

In Part 1 you will learn to adjust the micropipet. Look at the photographs, then follow this procedure:

1.1 Verify that the *Rate Controller* is in the *middle* of its range (Figure 2-1).

1.2 Verify that the wings are *unlocked* (Figure 2-2).

1.3 Set the volume—this is the *most important step—read these details two or three times* until you are sure you understand. Always *watch the volume window* (Figure 2-3) *while turning the top knob* (Figure 2-1) to adjust the volume.

Never adjust the volume above 200 µL or below 002 µL.

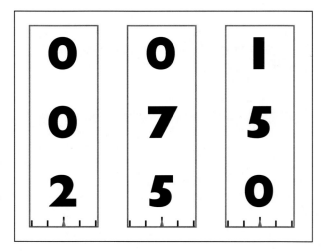

FIGURE 2-3
The volume indicator is read from the top down. Left: 2 µ L; middle: 75 µ L; right: 150 µ L. **WARNING: Never set volume above 200 µL or below 002 µL!**

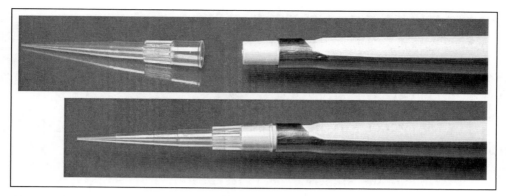

FIGURE 2-4
Attach a disposable tip with firm pressure.

For example, to set the volume for 150.0 µL, turn the top knob so the volume window reads about 150. Then turn the knob about 1/3 of a turn higher than 150. Then slowly turn the knob and stop at exactly 150.0.[1]

1.4   Firmly *attach a disposable tip.* Do not touch the end near the narrow opening (Figure 2-4).

1.5   *Press the plunger* (Figure 2-1). You will hear a "click" as a magnet latches the plunger into its precise position. Remove your thumb from the plunger.

1.6   Hold the pipet nearly vertical, and place the *tip just below the surface* of the solution to be drawn into the pipet (Figure 2-5a).

1.7   *Press the trigger* to draw the solution into the tip. *Pause briefly* for the tip to fill, keeping the tip below the surface of the liquid.

1.8   *Withdraw the pipet* from the solution. Look for air bubbles in the tip or liquid on the outside. If there is a bubble, you need to start over. Wipe off any liquid outside with a KimWipe. Do not touch the tip opening.

1.9   *Dispense by touching the tip* against the inner wall of the receiving tube, just *above the surface* of any liquid in the tube (Figure 2-5b). Then press the plunger to the bottom. Wait 1–2 seconds with the tip touching the tube wall.

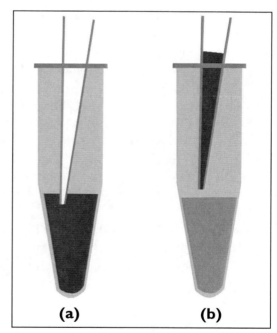

(a)   **(b)**

FIGURE 2-5 (a) Place the end of the tip 2–3 mm below the surface in the tube containing the solution to be dispensed. (b) Touch the tip to the wall of the receiving tube just above the surface of any liquid in the tube.

1.10 *Press the tip ejector* to discard the tip. Use a new tip, when necessary, especially to avoid contaminating stock solutions.

1.11 Repeat steps 1.3–1.10 as needed.

---

[1]Dialing down toward the desired volume eliminates systematic "backlash" errors in the internal mechanism of the pipet.

2.   Practicing with the Micropipet and Vortex Mixer

In Part 2 you will practice using the pipet. Your instructor will provide snap-top tubes, a micropipet, tips, a labeling pen, a bottle of water, and a bottle of a colored solution. You will dilute the colored solution with water to a final volume of 200.0 µL.

2.1   One partner does the work through step 2.10; the second partner starts at step 2.11. Both partners should observe. <u>Record answers to the underlined queries and any other observations on the Notes pages at the end of this module.</u> Your instructor will sign your Notes at the end of the lab period and distribute a report sheet. You will turn in your completed report at the beginning of the next period. Talk to your instructor if you encounter a problem or have a question.

2.2   Use a squeeze bottle or a transfer pipet, not a micropipet, to put about 1 mL of water into a snap-top microcentrifuge tube (Figure 2-6). Label the tube as "1" and place it in a support rack. A transfer pipet looks like an eyedropper.

2.3   Use a squeeze bottle to put about 1 mL of the colored solution into another snap-top. Label this as tube "2."

2.4   Load 100.0 µL of water into a micropipet tip, following the instructions from 1.3–1.8. The tips have mold marks that help to confirm that you have drawn the right amount of liquid into the tip – which mark corresponds to 100 µL? Get used to checking these marks.

2.5   Dispense the water into a fresh snap-top, labeled "100." The same tip can be used for water throughout this part of the module. Later, you will add the proper volume of the colored solution.

2.6   Use the same tip to dispense 175.0 µL of water into another tube, labeled "25." Remove the tip and lay it on a KimWipe.

2.7   Load 25.0 µL of the colored solution into a new pipet tip. Check the mold mark on the tip.

2.8   Add the colored solution to "25." Snap the top closed tight.

2.9   Mix the contents of "25" for 2–3 seconds on the vortex mixer. Then set it in the rack. Your instructor will give you directions on how to use the mixer.

2.10 Repeat steps 2.7–2.9, adding 100.0 µL of colored solution, to "100." Check the mark.

2.11 The second partner should repeat steps 2.7–2.9, dispensing 150.0 and 190.0 µL of water, and then adding 50.0 and 10.0 µL of colored solution to make a total of 200.0 µL. Label each tube appropriately. Check the mold marks and begin to develop a sense of how full the tip should be when it contains a certain volume.

2.12 <u>Rank the four samples "10" to "100" in order of increasing dilution.</u> Both partners should record the results. In general, "more dilute" means "less concentrated."

2.13 Dispose of the contents of your tubes according to your instructor's directions.

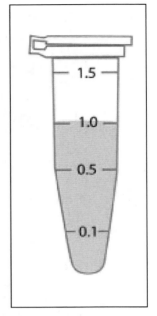

FIGURE 2-6
A snap-top microcentrifuge tube containing 1 mL.

3.   Milligram Balance

A molecular formula can be used to determine the number of moles of molecules or atoms in a substance from its mass. Many research labs have balances that can weigh to ± 1 μg (0.000001 g), using special precautions to eliminate air currents and other sources of error. Such balances can be used to measure small masses and to determine larger amounts with five significant figures or more. In this laboratory, we will use balances that are accurate to *± 1 mg (0.001 g)*. Even at this level of accuracy, contributions from *air currents*, the moisture in *fingerprints*, and *evaporation* from aqueous solutions can *affect the last digit* in the measurement.

In Part 3, you will weigh different calibrated masses to check the accuracy of the balance. We have a limited number of balances, so aim to finish Part 3 in five minutes, then go on to Part 4. One partner does the work in Part 3 and the other does the work in Part 4. Both partners should record observations and answers to the underlined queries in his/her own Notes.

Balances are delicate, so treat them with respect.

Do not use for any mass greater than 100 g.

Never drop anything heavy on the pan.

Clean up spills immediately.

Report any unusual circumstances to your instructor.

3.1   Place a small plastic weighing boat on the pan of the balance. Wait a few seconds and press "Tare." The balance should read 0.000 ± 0.001.

3.2   Vigorously wave your hand above the balance. How much does the reading change while you are waving? Do you think air currents will be a random or a systematic source of errors in weighing samples? Why?

3.3   Obtain a set of weights from your instructor. Use forceps to transfer individual weights to the weighing boat. Weigh and record two masses between 100 mg and 50 g. Reset the tare before each measurement.

3.4   Weigh and record the mass of a dry snap-top tube.

3.5   Pick up the same tube. Dip your fingertips in water and shake off the excess. Hold the snap-top in your wet fingers for a few seconds, and set it back in the weighing boat. Record the new mass. How much mass was gained? Will excessive moisture on your fingers cause an error in weighing? If so, will the error be systematic or random? Why?

4. Calibrating the Micropipet

In this part of the module you will weigh several different volumes dispensed from the micropipet.

4.1   Aim to finish in 5–10 minutes.

4.2   Review steps 1.1 to 1.9, from Part 1. Do NOT discard the tip between measurements.

4.3   Set pipet to 150.0 µL, as in the first section.

4.4   Tare the balance with a clean, dry weighing boat.

4.5   Transfer 150.0 µL of water directly to the weighing boat. <u>Record the mass</u> as quickly as possible. <u>Record the mass again after 60 seconds</u>. <u>Does it change? Why? Would it change more rapidly on a dry or rainy day? Why?</u>

4.6   Tare the balance with a clean dry weighing boat.

4.7   Transfer 50.0 µL of water to the weighing boat. <u>Record the mass</u> as quickly as possible.

4.8   Tare the balance and add a second 50.0 µL of water to the weighing boat. <u>Record the mass</u> as quickly as possible.

4.9   Tare the balance and add a third 50.0 µL of water to the weighing boat. <u>Record the mass</u> as quickly as possible.

4.10 Return the weights to your instructor.

5. Cleanup and Check in

When you have finished all parts of the module:

5.1   Clean up the balance area and your work area. Dispose of tips and boats according to your instructor's directions.

5.2   Return the pipettor and other supplies.

5.3   Write the <u>temperature of the room</u> in your Notes.

5.4   *Clearly print* your <u>lab partner's name</u> on your Notes report sheet.

5.5   Have your <u>instructor sign and date</u> your Notes.

5.6   Exchange contact information (email, phone) with your partner in case you have questions during the week.

5.7   Do not copy your report from your partner. Collaboration is encouraged, copying calculations or answers is not.

Temp (°F)	Temp (°C)	Density (g/mL)
60.8	16.0	0.9989
62.6	17.0	0.9988
64.4	18.0	0.9986
66.2	19.0	0.9984
68.0	20.0	0.9982
69.8	21.0	0.9980
71.6	22.0	0.9978
73.4	23.0	0.9975
75.2	24.0	0.9973
77.0	25.0	0.9970
78.8	26.0	0.9968
80.6	27.0	0.9965
82.4	28.0	0.9962
84.2	29.0	0.9959
86.0	30.0	0.9957

6. **Report**
    - The report is due at the *beginning* of the next lab period. Reports turned in after the quiz begins will be marked late. Reports will be marked down 5% for each day that they are late.
    - Be sure your name, your TA's name, and your section number are on the front of your report.
    - Be sure your partner's name is clearly printed on the page where your report begins.

6.1 In section 2, does the color fade in proportion to the extent of dilution with water?

6.2 Briefly explain the difference between random and systematic errors.

6.3 In sections 3 and 4, which of the following are random, and which are systematic errors? Explain each answer briefly.
    - Moisture in fingerprints
    - Air currents
    - Improperly calibrated balance
    - Improperly calibrated micropipet
    - Evaporation from a measured volume of solution

6.4 We often make the assumption that the density of water is 1.0000 g/mL (density is defined as mass per unit of volume). Look at the accompanying table.
    - To one significant digit, what percent error is introduced by this assumption in a room at temperature = 22°C? Show your calculation.
      Is this a random or systematic error? Explain briefly.

6.5 Analyze the three measurements of mass when 50.0 μL of water was added to a weighing boat (4.7 to 4.9). You don't need to correct for the density of water.
    - Compute the average volume that was delivered.
    - Compute the average deviation from 50.0 μL.
    - Compute the average percent error.

6.6 Does it appear that there is a systematic error greater than 5% in the calibration of your micropipet at any of the set volumes (150.0 or 50.0 μL)?

Branden Breazeale   Morgan Conover

Notes

.998 g/mL

$$98 \, mg \times \frac{1g}{1000mg} \times \frac{1000 \, mL}{1mL} = 98.2 \, \mu L$$

72.8°F

Least Dilute ⟶ Most Dilute

2, 100, 50, 25, 10, 1

Waving vigorously greatly altered the reading on the scale. It is a random error, because air currents fluctuate.

1.) 10.009 g

2.) 1.005 g

3.) Pop-top tube = 1.035 g

4.) Pop-top tube after water on fingertips = 1.038 g

0.003 g was gained from moisture on my hands. It's a systematic error, as precautions should be taken to prevent wet fingertips.

150 μL = 0.028 g
water

After 60 s   0.031 g

50 μL 1.) 0.089 g

50 μL 2.) 0.102 g

50 μL 3.) 0.151 g

Yes, slightly. Perhaps because of condensation/evaporation. It'd happen more quickly on a moist day, because it would gather moisture more quickly.

Notes

Laboratory 3

Chemical Formulas and Reactions

This module has two main intentions. The first is to describe the formulas for common ions and chemical compounds. The second is to observe several chemical reactions and describe their chemical transformations.

I. REVIEW OF CHEMICAL FORMULAS, STRUCTURES, AND ACID-BASE REACTIONS

Chemical formulas show the proportions at which atoms combine to make molecules. Balanced chemical equations extend this notation to show how atoms, ions, or molecules in substances rearrange to make different substances.

Important Note: Today's quiz will be on chemical nomenclature, so practice naming compounds BEFORE coming to lab. Success in General Chemistry requires that you know the chemical shorthand. If your skills are weak in remembering chemical formulas, it is strongly recommended that you buy a pack of blank index cards, write the names listed in Tables 1 and 2 on one side and the formulas on the other. Practice several times daily until you know them, then review before each exam or quiz. A cation (pronounced, *cat-eye-on*) carries a positive charge in water under most laboratory conditions, and anions (*ann-eye-on*) are negatively charged. Cations and anions combine to form chemical substances referred to as salts, where the charges neutralize each other. Many dry solid salts, like sodium chloride, dissolve readily in water to give cations and anions separated from each other by solvent molecules. Other salts, like barium sulfate, are very insoluble in water.

As you read this section, refer to the periodic table of the elements at the end of this book. The periodic chart is the key to understanding and predicting the ion-forming and bond-forming characteristics of atoms.

1.1 Common Cations

The formulas and names of many other ions can be inferred from those listed in Table 1. Atoms in vertical columns of the periodic chart usually have very similar chemical properties. The progression of cations going down Table 1 moves across the periodic chart from left to right. Look

Table 1.   The Most Common Ions (Memorize this list)

CATIONS		ANIONS	
Hydrogen	H^+	Hydride	H^-
Lithium	Li^+	Fluoride	F^-
Sodium	Na^+	Chloride	Cl^-
Potassium	K^+	Perchlorate	ClO_4^-
Magnesium	Mg^{2+}	Hypochlorite	ClO^-
Calcium	Ca^{2+}	Chlorate	ClO_3^-
Barium	Ba^{2+}	Bromide	Br^-
Chromium(III)	Cr^{3+}	Iodide	I^-
Manganese(II)	Mn^{2+}	Oxide	O^{2-}
Iron(II)	Fe^{2+}	Hydroxide	OH^-
Iron(III)	Fe^{3+}	Peroxide	O_2^{2-}
Cobalt(II)	Co^{2+}	Sulfide	S^{2-}
Nickel(II)	Ni^{2+}	Sulfate	SO_4^{2-}
Copper(II)	Cu^{2+}	Hydrogen sulfate	HSO_4^-
Silver	Ag^+	Nitrate	NO_3^-
Gold(II)	Au^{2+}	Phosphate	PO_4^{3-}
Zinc	Zn^{2+}	Hydrogen phosphate	HPO_4^{2-}
Mercury(II)	Hg^{2+}	Dihydrogen phosphate	$H_2PO_4^-$
Aluminum	Al^{3+}	Carbonate	CO_3^{2-}
Tin(II)	Sn^{2+}	Bicarbonate	HCO_3^-
Lead	Pb^{2+}	Cyanide	CN^-
Ammonium	NH_4^+	Acetate	$C_2H_3O_2^-$
		Chromate	CrO_4^{2-}
		Dichromate	$Cr_2O_7^{2-}$
		Permanganate	MnO_4^-

at the periodic chart and find strontium, Sr. Note that it lies between Ca and Ba, so from Table 1 you can infer that its most common ion is Sr^{2+}.

Some cation names in the table are followed by (II) or (III), which designates the charge on a monatomic ion. Including the Roman numeral means that these atoms have other common forms beyond those shown in Table 1. The Roman numeral is also referred to as the oxidation state[1] for these atoms. Oxidation states correlate with the number of electrons that are associated closely with the nucleus of the atom. There is no need to include the Roman numeral for most of the cations shown in the table as they rarely exist in other oxidation states. Most cations originate from substances that have the characteristics of metals.

[1]Oxidation states will be covered in detail later in the lectures.

**Table 2.   A Few Common Bases and Acids
(Memorize this list)**

BASES	
Sodium hydroxide	NaOH
Potassium hydroxide	KOH
Ammonia	NH_3 (aq)
Magnesium hydroxide	$Mg(OH)_2$
Calcium hydroxide	$Ca(OH)_2$
Iron(II) hydroxide	$Fe(OH)_2$
Iron(III) hydroxide	$Fe(OH)_3$
Aluminum hydroxide	$Al(OH)_3$
ACIDS	
Hydrochloric acid	HCl
Perchloric acid	$HClO_4$
Sulfuric acid	H_2SO_4
Nitric acid	HNO_3
Phosphoric acid	H_3PO_4
Acetic acid	$HC_2H_3O_2$
Carbon Dioxide	CO_2 (aq)
Hydrofluoric acid	HF
Hydrobromic acid	HBr
Hydroiodic acid	HI

1.2 Common Anions

The progression of anions going down Table 1 sweeps across the periodic chart in the opposite direction—from right to left.[2] In addition to the few monatomic anions, there is a wide array of polyatomic anions. Table 1 arranges them by the first main atom in their name. Again, one can often infer the formulas for other ions from Table 1 and the periodic chart. For instance, arsenic, As, is below phosphorus, P, in the periodic chart, and arsenate ion, AsO_4^{3-}, has a similar formula to PO_4^{3-}.

Carbon, nitrogen, phosphorous, silicon, and similar elements occupy a region of the periodic chart between the elements that tend to form monatomic cations or anions. C, N, P, Si, etc., tend to have a wide range of oxidation states and can form an uncountable array of covalent compounds. While covalent bonds will be covered later in the semester, at this point it is sufficient to say that these bonds are made from electrons that are shared between two or more atoms. Thus, for example, the collection of atoms in bicarbonate ion, HCO_3^-, is held together by covalent bonds; this aggregate has one extra electron, which gives its overall net charge of −1. Most metals can participate in similar bonds, which lead to ions like permanganate, with five atoms, or dichromate, with nine.

[2]Hydrogen has properties both of metals (monatomic cations) and nonmetals (monatomic anions, like chloride, oxide, etc.)

1.3 Acids, Bases, Salts, and Chemical Equations

The concept of acid-base reactions is one of the oldest systems for understanding chemical reactions. Although we strongly discourage the practice today, early chemists tasted everything: acids were distinguished as being sour (the German word for acid is Säuere), bases as bitter, and salts tasted salty. As you learn more about chemistry, you will find that the theory of acids and bases evolves to become more and more general. It is one of the most powerful tools to understand and predict chemical reactions.

There are only a few acids and bases that are found in nearly every chemistry and biochemistry laboratory on the planet. These are the ones shown in *italic* type in Table 2. Also included in italics is the acid, aqueous CO_2, which is rarely used as an acid by bench chemists. However, it is the most important source for regulating the acidity of living cells, a topic of great interest to biochemists and biologists. That makes it worth including among the most common laboratory acids and bases.

BASES

When a base is dissolved in water, a cation and anion are produced:

(1) $\quad NaOH \rightarrow Na^+ + OH^-$

(2) $\quad Ca(OH)_2 \rightarrow Ca^{2+} + 2\ OH^-$

(3) $NH_3 + H_2O \rightarrow NH_4^+ + OH^-$

All of the cations in Table 1 can be derived from a base in a reaction similar to that shown in reactions (1–3). The names of the common bases are usually the cation name followed by "hydroxide," like "sodium hydroxide."

ACIDS

When an acid is dissolved in water, again a cation and anion are produced:

(4) $\quad HCl \rightarrow H^+ + Cl^-$

(5) $\quad H_2SO_4 \rightarrow 2\ H^+ + SO_4^{2-}$

(6) $CO_2 + H_2O \rightarrow H^+ + HCO_3^-$

All of the anions in Table 1 can be derived from an acid in a reaction similar to that shown in (4–6). The names of the common acids must be memorized—they are similar but not always the same as the names of the anions.

All ions and most molecular substances in water are "aquated," that is, associated tightly with several water molecules. This will occasionally be noted specifically, as for CO_2 (aq) and NH_3 (aq) in Table 2. Older (and flawed) usage describes carbonic acid, H_2CO_3 ($H_2O + CO_2$), and ammonium hydroxide, NH_4OH ($NH_3 + H_2O$). The usage of H_2CO_3 and NH_4OH is a relic of one of the older definitions for acids and bases. The problem is that neither exists in appreciable concentration as a molecular species in water.

STRONG/WEAK VS. CONCENTRATED/DILUTE

Beginning students often confuse the strength of an acid or base with its concentration. "Strength" has to do with how fully the reactants in (1–6) move toward the products. All of the italicized bases in Table 2 are strong bases except for ammonia, and all of the italicized acids are strong acids except for carbon dioxide. Thus, in equation (3), ammonia produces little OH^- per mole of NH_3 that is dissolved. Likewise, in (6), little H^+ is produced per mole of CO_2 dissolved. On the other hand, a solution with 2 moles of the strong base, NaOH, per liter of water is less concentrated than a solution with 10 moles of the weak base, NH_3, per liter.

NEUTRALIZATION

One of the first chemical equations a student chemist learns is:

(7)   Acid + Base $\rightarrow$ Salt + Water

For instance:

(8) $HCl + NaOH \rightarrow NaCl + H_2O$

or

(9) $H^+ + Cl^- + Na^+ + OH^- \rightarrow Na^+ + Cl^- + H_2O$

In (9), NaOH has been replaced with the component ions from (1) and HCl with the ions from (4). Na^+ and Cl^- occur on both sides and can be canceled, just as in a mathematical equation. They are sometimes called "spectator" ions—watching the important acid-base reaction:

(9) $OH^- + H^+ \rightarrow H_2O$

Of course, there are exceptions to most rules, including (7). Consider the reaction of ammonia with carbon dioxide in water:

(10) $CO_2 + NH_3 + H_2O \rightarrow NH_4HCO_3.$

Water is a reactant, rather than a product as it is in most other acid-base reactions.

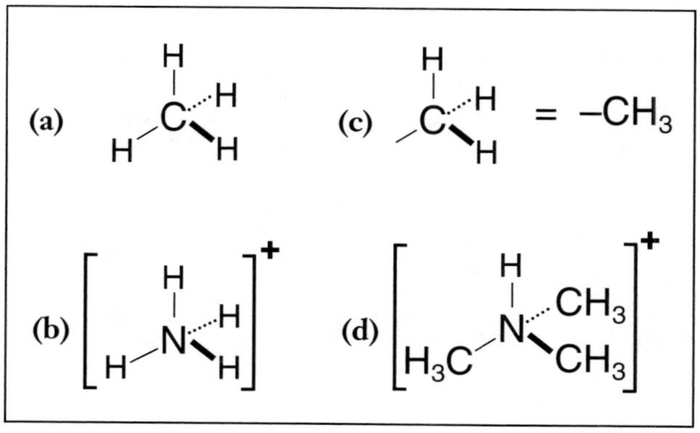

FIGURE 3-1
Comparison of the structures of (a) methane, (b) ammonium ion, (c) methyl group, and (c) trimethylammonium ion.

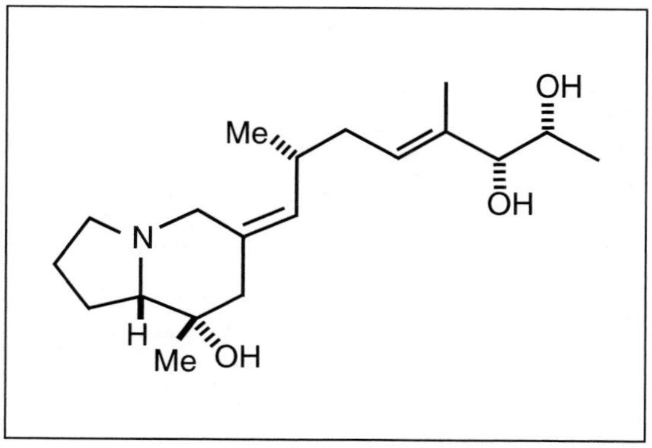

FIGURE 3-2
The compound, (+)-pumiliotoxin B, prepared by organic chemists.

BALANCED EQUATIONS

An important skill is to be able to balance chemical equations. Notice that each of the equations given is balanced—the number of atoms of each kind is the same on both sides, and the total charge balances, as well. Your text has lots of practice examples. If you are having problems, come to any of the office hours, which are listed at the course Web site.

1.4 Polyatomic Ions and Molecular Structure

Ammonium ion, NH_4^+ (Figure 3-1b), is a good example of a polyatomic ion. A larger relative is the $(CH_3)_3NH^+$, "trimethylammonium" ion, where three of the covalent N–H bonds of ammonium have been replaced by N–C bonds to "methyl" groups. Figure 3-1b and d emphasize the similarity between the "structures" of ammonium and trimethylammonium ions. As can be seen, chemists have developed a shorthand way of representing structures. For these simple structures, the atoms are represented by letters and bonds by lines. In three dimensions, dashed lines indicate that the atom or group is behind the plane of the page; heavy lines indicate that the atom or group is in front.

A more complicated structure is typical of compounds called, "natural products." Figure 3-2 illustrates a chemical that South American natives use for poison darts. They get the poison from frogs, but it can also be synthesized in the lab. The figure is simplified to show the essential components of the molecule: most hydrogen atoms are left out, carbons are assumed to be present at the points joined by bonds (lines), other atoms are named specifically, and "Me" is an abbreviation for a methyl group. An H-atom or other group of atoms is shown in front or behind the page if that is critical in understanding the properties of the molecule. The three dimensional shapes of these molecules can be difficult to represent on a page, although the shapes are often crucial to biological function and other properties. Molecules that focus mainly on the chemistry of carbon are placed in the domain of "organic chemistry." That contrasts with popular culture's definition of "organic" foods that are promoted as being especially healthy. Many poisons, pesticides, and preservatives are classified as organic chemicals as they have carbon frameworks. Many people have a generalized fear of "chemicals" in food when, in fact, food is entirely made up of chemicals. Furthermore, a human body consists of thousands of chemicals working in a coordinated fashion.

The domain of "inorganic chemistry" includes compounds where the most interesting aspect involves metal or other non-carbon atoms. Often their three-dimensional structures are complicated but beautiful, as in Figure 3-3. This structure has barium-oxygen bridging bonds and 3, 5, and 6-membered organic rings; the balls represent atoms (C = small, O = medium, Ba = large, and H = not shown).

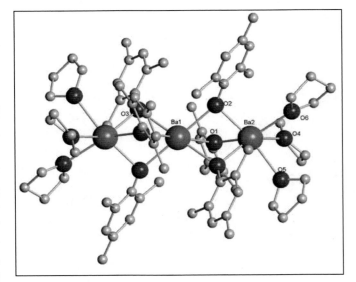

FIGURE 3-3
A compound prepared by inorganic chemists. Note the symmetry about the central atom, Ba1. Drawing by K. Ruhlandt.

The chemist's shorthand way of representing molecules gets ever more sophisticated in order to emphasize the most important and interesting aspects of a structure. For example, the nucleocapsid protein from HIV-1 (human immunodeficiency virus) is a polyatomic ion with a +9 charge; its chemical formula, $Zn_2C_{263}N_{84}H_{427}S_8O_{74}{}^{9+}$, doesn't show much that is useful. If we were to draw this structure with letters and lines as in Figures 3-1 to 3-2, it would look like a spaghetti mess of crossing lines and overlapping letters. The shorthand described earlier has evolved to allow an efficient description of such a structure (Figure 3-4).[3]

1.5 Hydrates

Ions in aqueous solution are tightly associated with water molecules. When many salts crystallize from aqueous solutions, they bring along some of the H_2O molecules. Often the water molecules occupy specific sites in the crystal lattice, giving rise to simple proportions between water, cations, and anions. For instance, copper sulfate is commonly available as $CuSO_4 \cdot 6H_2O$, where the dot "·" means that the simplest formula for the substance has six water molecules along with one Cu^{2+} and one $SO_4{}^{2-}$. We could also write the formula as $CuH_{12}SO_{10}$, but that would obscure the important chemical fact that the sulfate anion, $SO_4{}^{2-}$, is present, and that the crystal is of the salt, $CuSO_4$. The water associated with the crystal is often referred to as "water of hydration." The water molecules can often be driven off from the solid by heating; they are usually held much less strongly than the strong interactions between the ions in the crystal.

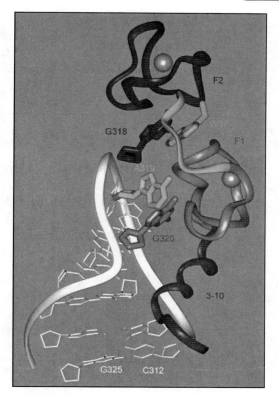

FIGURE 3-4
Three-dimensional structure of the HIV-1 nucleocapsid protein (above, right) interacting with the RNA genetic material (below, left). The ribbons show the approximate locations of the repeating protein or RNA main chain atoms. Certain side chains are highlighted, including guanine (318, 320) and adenine (319) bases, and the tryptophan (W37) amino acid.

1.6 Complexes

In addition to hydration, ions can interact with other materials. For instance, Cu^{2+} ion in solution makes a stable complex with NH_3: $Cu(NH_3)_4{}^{2+}$. Many such complexes are brightly colored.

Biochemical complexes are often _very_ complex. The example in Figure 3-4 shows the nucleocapsid protein complexed with a fragment of HIV's genetic material, RNA. The interaction between the highlighted RNA bases, G318 and G320, with the amino acids surrounding the spherical zinc atoms is primarily responsible for packaging HIV RNA into new viruses. A key to understanding complicated molecules is their three-dimensional structure, using drawings that

[3]A.C. Paoletti _et al._ (2002) _Biochemistry 51_, 15423-8; "Affinities of the nucleocapsid protein for variants of SL3 RNA in HIV-1."

leave out insignificant details, while emphasizing the most important features. Chemists, bio-chemists, and molecular biologists use these 3D structures to help understand how such molecules interact and how that relates to their chemical or biological function. Still, it is interesting that the most important building blocks of proteins are called amino *acids*, and those in DNA and RNA are called *bases*.

2. EXPERIMENTING WITH CHEMICAL REACTIONS

In Part 2, you will perform several reactions. Your instructor will tell you where the bottles of solutions are located. Use a different pipet tip for each solution. Your instructor also will tell you how to properly dispose of the chemicals after the reaction is done. Be careful that you do not contaminate the stock solutions. If you take too much of a solution, do not try to put it back in the bottle—put the excess in the waste disposal container. Notify your instructor if a bottle is empty or appears to have been contaminated.

You will be using dilute solutions of acids, bases, and caustic chemicals that are 10–100 times less potent than the most concentrated solutions. However, they can still cause burns if spilled on your skin or in your eyes. Report any spills to your instructor IMMEDIATELY. Avoid prolonged exposure to vapors. Be sure to follow the special directions given by your instructor regarding safety and disposal.

One partner does the work for the odd numbered experiments, the second partner does the work for the ones with even numbers. Both partners should observe and each should <u>record observations and answers to the underlined queries in his/her own Notes section</u>. Talk to your instructor if you encounter a problem or have a question. While doing the experiments, save a few blank lines in your Notes to write your balanced chemical equations and anything you may want to add later about the reaction.

2.1 <u>*BaCl₂ with H₂SO₄*</u>. Pipet 100 µL of $BaCl_2$ solution into a snap-top tube and place it in a support rack. Put ~1 mL of H_2SO_4 solution into a different snap top. Label each tube.

Now, pipet 100 µL of the H_2SO_4 solution into the $BaCl_2$. <u>Record what happens</u>. <u>Write a balanced chemical equation to describe the change that occurs</u>.

Keep the remaining H_2SO_4 solution for several parts yet to come, and dispose of the $BaCl_2/H_2SO_4$ mixture.

2.2 <u>*NaOH with H₂SO₄*</u>. Obtain ~1 mL of NaOH solution in a snap top. Put a drop of phenolphthalein in a second tube, and get a ~1 inch strip of pH paper. Label each tube. The number of moles of H_2SO_4 (from 2.1) and NaOH are nearly the same per mL of the solutions. <u>Record what happens as you perform the following steps. Write a balanced chemical equation to describe the change that occurs</u>.

Add exactly 100 µL of the acid to the tube containing phenolphthalein. Add 100 µL of the base and mix. Add 80 µL more of the base and mix. Add 20 µL more base and mix. Repeat the addition of 20 µL more base.

Put 20 µL of the base on the pH paper and 20 µL of the acid on a different spot. <u>Record what happens</u>.

Keep the remaining NaOH solution for several parts yet to come, and dispose of the $NaOH/H_2SO_4$ mixture.

2.3   *NaOH with HC₂H₃O₂*. Obtain ~1 mL of $HC_2H_3O_2$ solution in a snap top. Put a drop of phenolphthalein in a second tube. Label each tube. The molar quantities of NaOH (from 2.2) and $HC_2H_3O_2$ are nearly the same per mL of the solutions. <u>Record what happens as you perform the following steps. Write a balanced chemical equation to describe the change that occurs.</u>

Add exactly 100 μL of the base to the tube containing phenolphthalein. Add 80 μL of the acid and mix. Add 20 μL more of the acid and mix. Repeat the addition of 20 μL more acid.

Put 20 μL of the acid on the pH paper. <u>Record what happens.</u>

Dispose of the $NaOH/HC_2H_3O_2$ mixture.

2.4   *NH₃ with H₂SO₄*. Obtain ~1 mL of aqueous ammonia solution in a snap top. Put a drop of phenolphthalein in a second tube. Label each tube. The molar quantities of H_2SO_4 (from 2.1) and NH_3 are nearly the same per mL of the solutions. <u>Record what happens as you perform the following steps. Write a balanced chemical equation to describe the change that occurs.</u>

Open the top of the tube containing the ammonia solution, and waft the odors toward your nose as instructed in the module, "For Your Safety." <u>Record</u> your observations.

Add exactly 100 μL of the ammonia solution to the tube containing phenolphthalein. Add 30 μL of the acid and mix. Add 10 μL more of the acid and mix. Add 10 μL more acid and mix. Repeat the addition of 10 μL more acid.

Put 20 μL of the ammonia solution on the pH paper. <u>Record what happens.</u>

Open the tube containing the stock NH_3 solution. Moisten the end of another strip of pH paper, and hold above the open tube, but do not touch the solution. <u>Record</u> your observations.

Keep the remaining ammonia solution for several parts yet to come, and dispose of the NH_3/H_2SO_4 mixture.

2.5   *FeCl₃ with NaOH*. Obtain ~0.5 mL of $FeCl_3$ solution in a snap top. <u>Record what happens as you perform the following steps. Write a balanced chemical equation to describe the change that occurs.</u>

Pipet 100 μL of $FeCl_3$ solution into a snap-top tube and place it in a support rack. Label the tube.

Pipet 100 μL of the NaOH solution into the $FeCl_3$ and mix. Centrifuge for a minute. Add 100 μL more of the NaOH and observe what happens. Repeat the mix and spin steps, then add another 100 μL of NaOH and observe.

Dispose of the $NaOH/FeCl_3$ mixture.

2.6   *Mg with H₂SO₄ and HC₂H₃O₂*. Obtain two small pieces of magnesium in two snap tops. <u>Record what happens as you perform the following steps. Write balanced chemical equations to describe the changes that occur.</u>

Pipet 100 μL of H_2SO_4 solution into the first snap-top tube with Mg and <u>note</u> what happens over 15 minutes. Label the tube.

Pipet 100 μL of $HC_2H_3O_2$ solution into the second snap-top with Mg and <u>note</u> what happens over 15 minutes. Label the tube.

Dispose of the contents after 15 min of observation.

2.7 _CuSO₄ with nail_. Obtain ~0.5 mL of $CuSO_4$ solution in a snap top. Obtain a small nail. <u>Record what happens as you perform the following steps. Write balanced chemical equations to describe the changes that occur.</u>

Pipet 200 μL of $CuSO_4$ solution into a snap-top tube. Add the nail, but don't mix, and <u>note</u> what happens over 15 minutes. Dispose of the tube and contents.

2.8 _NaOH with NH₄HCO₃_. Use a metal spatula to obtain NH_4HCO_3 crystals in two snap tops—about half-way to the 0.1 mL mark is sufficient. <u>Record what happens as you perform the following steps. Write a balanced chemical equation to describe the change that occurs.</u>

Open the top of one tube containing the crystals, and waft the odors toward your nose. <u>Record</u> your observations.

In the open tube, pipet 200 μL of NaOH solution onto the ammonium bicarbonate crystals, close the top, and mix to dissolve. Label the tube. Hold the tube away from your face, and snap open the top. Waft the odors toward your nose. <u>Record</u> your observations.

Heat the tube to 60°C in a heating block with the top open (2–3 min). Waft the odors toward your nose. <u>Record</u> your observations.

Dispose of the solutions used in this part.

2.9 _NH₄HCO₃ with heat_. <u>Record what happens as you perform the following steps. Write a balanced chemical equation to describe the change that occurs.</u>

Place the second snap top from the previous step in the 60°C heating block with the lid open (2–3 min). Waft the odors toward your nose. <u>Record</u> your observations.

Dispose of the materials used in this part.

3. CLEANUP AND CHECK IN

When you have finished all parts of the module:

3.1 Clean up the balance area and your work area. Dispose of liquid waste, tips, and tubes according to your instructor's directions.

3.2 _Clearly print_ your <u>lab partner's name</u> in your Notes and report. Exchange contact information (email, phone) with your partner—you may have questions during the week. Do not copy your report from your partner. Collaboration is encouraged, copying is not.

3.3 Return the pipettor and other supplies to your instructor. Have <u>instructor sign and date</u> your Notes.

4. REPORT

- Reports are due at the _beginning_ of the next lab period.
- Be sure your name, your TA's name, and your section number are on the front of your report.
- Be sure your partner's name is clearly printed on the page where your report begins.
- Be sure you have written answers to each underlined query in section 2 of the experiment.

Notes

2.1) Turned cloudy, milky white color immediately

$$BaCl_2 + H_2SO_4 \rightarrow BaSO_4 + 2 HCl$$

2.2) Colorless w/ acid + phen; when base was added, it became pink, then immediately back to colorless.

80μL, bright pink, then became colorless again. 20μL more = goes pink, stays pink. 20μL more, darker pink.

On pH paper, acid yielded an orange color, base yielded a blue color. $2NaOH + H_2SO_4 \rightarrow Na_2SO_4 + H_2O$

2.3) When 100μL base was added, the phen. turned deep pink.

When 80μL of acetic acid was added, there was no color change.

When 20μL more of acetic acid was added, bottom of solution became clear. $NaOH + HC_2H_3O_2 \rightarrow NaC_2H_3O_2 + HOH$

When 20μL more of acetic acid was added, the entire solution became clear.

When acid was put on pH paper, it turned a bright orange.

2.4) We cannot detect a smell from the NH_3

When 100μL of NH_3 was added to the phenolp, it became pink.

When 30μL H_2SO_4 was added to NH_3 solution, it became pale pink. $NH_3 + H_2SO_4 \rightarrow (NH_4)_2SO_4 + H_2$

When 10μL H_2SO_4 was added to solution, there was no change.

When 10μL H_2SO_4 was added to solution, it became clear again.

Finally, when 10μL more of H_2SO_4 was added, it remained clear.

When moist pH paper was held above NH_3, it became blue.

2.5) When 100μL of $NaOH$ is added to the $FeCl_3$, it becomes a darker yellow/orange. $FeCl_3 + NaOH \rightarrow Fe(OH)_3 + NaCl$

When 100μL more of $NaOH$ is added to solution, the bottom becomes an amber color.

(After centrifuge, the bottom has become brown, the rest of solution is completely clear.

2.6) 12:17, added 100μL H_2SO_4 to Mg chip. immediately began dissolving. $Mg + H_2SO_4 \rightarrow MgSO_4 + H_2$

at 12:32, we see a decrease in chip size.

12:19, added 100μL of acetic acid to Mg chip, immediately started dissolving.

at 12:34, there are bubbles, but the Mg chip appears unscathed. $Mg + HC_2H_3O_2 \rightarrow MgC_2H_3O_2 + H_2$

2.7) 12:24 iron nail is set in $CuSO_4$, appears to begin rusting.

12:37, the part of the iron nail in solution appears to be completely rusted.

$$CuSO_4 + Fe(III) \rightarrow Fe_2(SO_4)_3 + Cu$$

$\longrightarrow$ over

Notes

2.8) upon wafting, we discover ~~smell~~ from crystals.
    no smell.

Upon adding 200 μL of NaOH to crystals, the scent has become pungent.

After heating the NaOH with crystals to 60°C, the smell wasn't as strong, but it remained.

$$NaOH + NH_4HCO_3 \longrightarrow NaHCO_3 + NH_4 ~~\text{(crossed out)}~~ + H_2O$$

2.9) After heating the crystals, the smell from it is very very strong.

$$NH_4HCO_3 + heat \longrightarrow NH_4 + ~~\text{(crossed out)}~~ CO_2 + H_2O$$

Laboratory 4

The Acetic Acid Content of Vinegar

TIME	About 2 hours
MATERIALS	Commercial colorless vinegars, burettes, Erlenmeyer flasks, phenolphthalein indicator solution, standard sodium hydroxide solution (0.05 g of NaOH per mL of solution)

I. SAFETY AND CHEMICAL HAZARD GUIDE

Please review this before performing the experiment.

1.1 You must wear glasses at *all* times when in the laboratory!

1.2 Vinegar. Avoid splashing into eyes or on open cuts. Although vinegar presents no special hazards, more concentrated solutions of acetic acid should be handled with extreme caution. Concentrated solutions of acetic acid can cause asphyxiation, nausea, vomiting, chest pains, weakness, stupor, convulsions, and eye irritations. Avoid inhalation, ingestion, and skin contact.

1.3 Sodium hydroxide (1.25 M). CAUTION. This chemical can cause skin burns on contact. Wash affected area immediately with plenty of water, and notify your instructor. It can cause eye, nose, and throat burns, and temporary loss of hair on the skin, upon contact.

1.4 Phenolphthalein indicator solution. Usually available as a 1 percent solution in ethanol. Avoid ingestion, inhalation, and skin contact. Phenolphthalein has laxative properties when ingested.

2. INTRODUCTION

Many of the products we use around our homes contain a wide variety of chemicals. Not only cleaning materials, clothing, and drugs but also food products contain chemicals. The company that produces a certain consumer item must make sure that the proper amount of each ingredient has been added; otherwise the product may not function properly. Adding too much or too little of a chemical could produce health problems, especially if the product happens to be vitamin pills or pharmaceuticals. Companies are concerned not only about the health factor but about the uniformity of their products; they want you, the consumer, to know what you are getting each time you buy the products. You certainly wouldn't want to buy a brand of ketchup that had a different color and a different taste each time you bought another bottle of it!

To ensure uniformity, most companies have quality control laboratories. In these laboratories, chemists and technicians check samples of each product to be sure that the product meets rigid specifications. Regardless of whether the product is perfume, toothpaste, vitamin pills, or processed food items, quality control personnel check the main chemical ingredients.

Vinegar is a household staple that you undoubtedly put on your salads and use to season many other dishes. It consists mostly of water, along with acetic acid, salt, herbs, and spices. Acetic acid is the substance that gives vinegar its characteristic taste and odor. It is very important that the amount of acetic acid in vinegar be kept around 4 to 5 percent by weight. The Food and Drug Administration requires that the acetic acid content must be at least 4 percent, but if the acetic acid content gets too high, the vinegar won't taste good. In this experiment you will act as a quality control chemist. Your task is to check the acetic acid content of vinegar. You will do this by analyzing samples of vinegar from your neighborhood supermarket. The analysis involves performing a titration. Before you try, your instructor will demonstrate the procedure.

3. DISCUSSION

Titration analysis is one of the most important and useful types of analysis in the field of analytical chemistry. The material to be analyzed is usually in solution and is allowed to react with a reagent solution the concentration of which is known. The analyst gradually adds the reagent solution to the sample from a burette. A burette is a measuring device that enables one to add the reagent solution slowly, until it just reacts with all of the constituent in the solution being analyzed. When this point—called the *equivalence point* of the titration—is reached, the analyst stops adding reagent solution. How is the equivalence point recognized? The answer lies in a substance called an *indicator*, which is usually placed in the solution being analyzed before the titration begins. The indicator does not interfere with the analysis; however, at the end of the titration, the indicator signals the equivalence point by turning a color (or changing from its initial color). From the burette, the analyst reads the volume of the reagent solution that is used to react with the constituent in the sample. By finding out the volume of reagent solution required, the analyst can determine the amount of constituent in the solution being analyzed.

The acetic acid content of vinegar is your subject for analysis by titration. In this experiment, the reagent solution contains sodium hydroxide of known concentration, which reacts with the acetic acid portion of the vinegar. The analyst adds the sodium hydroxide solution slowly from a burette until all the acetic acid has been neutralized. A chemical indicator called *phenolphthalein*, which is colorless in acid solutions and pink in basic solutions, signals the endpoint of the titration. In the untitrated sample, the phenolphthalein is colorless. At the endpoint of the titration, the phenolphthalein turns pink.

One can determine the percent acetic acid in the vinegar by using the following information:

a. The volume of NaOH used to titrate the sample.
b. The concentration of NaOH solution in grams of NaOH per milliliter of NaOH solution.

c.   The stoichiometry of the reaction,

$$NaOH + HC_2H_3O_2 \longrightarrow H_2O + NaC_2H_3O_2$$

Molecular mass = 40.0   Molecular mass = 60.0

which tells us that 40.0 g of NaOH react with 60.0 g of $HC_2H_3O_2$. In other words, the ratio in which these two substances react is

$$\frac{NaOH}{HC_2H_3O_2} = \frac{40.0\ g}{60.0\ g} \quad or \quad \frac{4.00\ g}{6.00\ g} \quad or \quad \frac{1.00\ g}{1.50\ g}$$

Later in the experiment, you will see why this information is important.

d.   The weight of your vinegar sample.

4. PROCEDURE

To simplify matters, we shall assume in this experiment that the density of vinegar is 1 g/mL. So when you measure the volume of your vinegar sample in milliliters, that will also be its mass in grams.

4.1   Rinse two burettes first with detergent solution and then with water. Test for a clean burette by filling the burette with water, then letting the water run out, and seeing whether or not any drops of water adhere to the walls of the burette. (No water should adhere to the burette.)

4.2   Rinse one of the clean burettes with a *small* amount (5 mL) of vinegar solution. Allow 1 mL to drain out the tip, and then, while rotating the burette slowly, pour the remainder out the top. Fill the burette with vinegar. Be sure that the tip of the burette is also filled and that there are no air bubbles.

4.3   Rinse the other clean burette with a *small* amount of the standard sodium hydroxide solution that you are going to use in this analysis. Pour this rinse out and then fill the burette with the standard sodium hydroxide solution (see Figure 4.1). Again, be sure that the tip is filled and that there are no air bubbles.

4.4   Obtain a clean 125-mL Erlenmeyer flask and allow exactly 25.0 mL of vinegar to flow into the flask from the burette.

Take all burette readings to the nearest 0.1 mL. Note that a burette reads in a direction opposite to that of a graduated cylinder. In other words, 0.0 is at the top and 50.0 is at the bottom. (Read the volume of the liquid at the bottom of the meniscus.)

4.5   You now have a 25.0-g sample of vinegar in your flask. Record its mass on the notes page. If by chance you allowed a little more or a little less vinegar to flow into the flask, that's okay. However, record the proper mass of your vinegar on the notes page.

4.6   Add about 25 mL of distilled water to the sample in the flask. This water does not affect your sample or the titration; it just serves to wash the sides of the flask and make the endpoint easier to see.

4.7   Add two or three drops of phenolphthalein indicator solution to the flask.

4.8   Record the initial burette reading of the NaOH burette and then titrate your vinegar sample with the sodium hydroxide solution. The signal for the end-point of the titration is a change in the color of the solution from colorless to pink. Record the final burette reading of the NaOH and determine the number of milliliters of sodium hydroxide solution used.

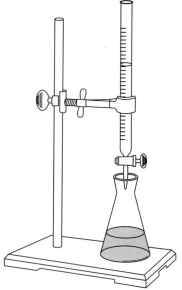

FIGURE 4-1
Burette filled with sodium hydroxide solution, clamped into position, and ready to titrate vinegar solution in Erlenmeyer flask. Many laboratories use a 3-finger burette clamp instead of the one shown here.

Proceed slowly. You do not want to overshoot the endpoint. With proper technique, one drop of NaOH solution should change the sample solution from colorless to pink as you reach the end point. If you do overshoot the endpoint, you can still salvage your sample. Here's how:

- Add additional vinegar to your flask from the vinegar burette until the phenolphthalein turns colorless. Swirl the flask to mix the contents. The mass of your vinegar sample has now been increased by the amount of vinegar you added. Be sure to note this change on the notes page when calculating the mass of the vinegar sample.

- Continue to titrate your vinegar sample with the sodium hydroxide. This time, try not to overshoot the endpoint. Record the final burette reading of the NaOH and determine the number of milliliters of NaOH solution used.

4.9 Perform two more titrations, each time using a fresh sample of approximately the same size as your first sample. Record all data on the notes page.

4.10 For each trial, calculate the percentage acetic acid in your vinegar sample.

5. CHEMICAL DISPOSAL GUIDE

- The titrated solutions and excess phenolphthalein indicator should be poured into a container marked **Aqueous Wastes.**
- Excess vinegar should be poured into a container marked **Acid Wastes.***
- Excess sodium hydroxide should be poured into a container marked **Alkaline Wastes.***

* Your instructor may have you neutralize these solutions prior to disposal.

Notes

$NaOH + HC_2H_3O_2 \rightarrow H_2O + NaC_2H_3O_2$

$mw = 40 g/mol$

$MW = 60 g/mol$

$\dfrac{40}{60} = \dfrac{NaOH}{CH_3COOH} = 1:.65$ ratio

Waste (3 bottles)

1. acetic acid
2. extra NaOH
3. aqueous waste from titration

grams NaOH

grams acetic acid

% acetic acid

g CH_3COOH neutralized

$0.85 g \text{ NaOH} \cdot \dfrac{1.5 g \, CH_3COOH}{1 g \, NaOH}$

$1.275 g \; CH_3COOH$

% CH_3COOH

$\dfrac{1.275 g \, CH_3COOH}{25 g \; vinegar} = 5.1\%$

mass vinegar	25.0 mL
initial volume NaOH	0.0
final volume NaOH	16.9
mL NaOH	16.9
[NaOH]	0.05 g mL

(Always)

$[NaOH] = 1.25 M$

amt. vinegar 25 mL = 25 g

amt. NaOH 17 mL

$\dfrac{1.25 mol}{1 L} \times \dfrac{40 g}{mol} = 50 g/L = 0.05 g/mL$ of NaOH

Mass vinegar	25 g
volume NaOH	0.0 mL
final volume NaOH	17.9 mL
mL NaOH	17.9 mL
NaOH	0.05 g mL

mass vinegar	25 g
initial volume NaOH	18.0 mL
final volume NaOH	35.8 mL
mL NaOH	17.8 mL
NaOH	0.05 g mL

CRYO 09/27/12

41

Notes

Laboratory 5
Calories from Food

1. BACKGROUND

The measurement of the energy content of a substance can be done by determination of the heat produced by it during combustion. During the combustion process, energy is released in the form of heat. The measurement of heat produced during a combustion process can be carried out in a calorimeter. A calorimeter is a device used to measure heat produced or absorbed during a process such as chemical reactions or combustion. The SI unit for energy is the joule; however, the unit of calorie is the one most commonly used to express energy measurements, particularly in foodstuff. A calorie is the measurement of the heat required to elevate the temperature of 1 g of water by 1°C. It is common to see the measurement of reactions in kilocalories (1 kcal = 1,000 cal). In the labels of foodstuff it is often represented as Cal to indicate kcal with the capital C indicating food calories.

A calorimeter is a device that is under constant pressure and thus by definition, the heat released (q) or measured is the same as the enthalpy or energy. Heat can then be calculated by the following equation:

$$q = \text{specific heat capacity} \times \text{mass} \times \Delta T$$

So in calorimetric measurements the change of temperature in the environment, in this case water, is measured directly and the heat released ($q_{environment}$) is calculated. For ease of understanding we assume that all of the heat released (q_{system}) by the combustion process is absorbed by the water in the calorimeter ($q_{environment}$). The specific heat capacity for water is 1.0 Cal/g°C and its density is 1.00 g/mL.

$$q_{system} = -q_{environment} \text{ will be noted as q(sys)}$$

2. OVERVIEW

In this experiment you will determine the calorie content in fat from peanuts, pecans, and other nuts by the use of a simple calorimeter. Under the experimental condition the combustion of the nuts will be incomplete. It is to say that not all of the nut material will be consumed for it is only the oil or fat that burns with ease. Therefore the heat released that we will be measuring is that from the combustion of the oil. The results of the experiment can be compared against the nutritional information of the food labels for calorie measurement from fat.

Calculation Example

A peanut with a mass of 0.500 g is burned and 0.125 g remains after burning (thus 0.375 g of nut burned). This produced a temperature change from 25.3°C to 32.8°C in 200 g of water.

The indicated serving size for a package of peanuts is 28 g; how many Calories (kcal) are in a serving?

Calculate the q_{water} using the change in temperature and the specific heat of water

$$q_{water} = \text{(specific heat of water)(mass of water)}(\Delta T)$$

$$q_{water} = 1.0 \text{ cal/g°C})(200g)(32.8°C-25.3°C)$$

$$q_{water} = 1500 \text{ cal} = 1.50 \text{ kcal} = 1.50 \text{ Cal}$$

Assume the amount of heat gained by the water is equal to the amount of heat released by burning the peanut.

$$\frac{Cal_{serving}}{28g} = \frac{1.50 \text{ Cal}}{0.375g}$$

$$Cal_{serving} = [(1.50 \text{ Cal})(28g)]/0.375g$$

$$Cal_{serving} = 112 \text{ Cal}$$

3. SAFETY

Goggles and closed-toe shoes are **always** required for any experiment in the laboratory.

CHEMICAL ALERT:

Peanuts   No hazard

Note: Let your instructor know if you have a known food allergy to peanuts or other nuts.

4. EXPERIMENT

You will need a thermometer, hot plate, funnel, soda can, and 400- or 600-mL beaker.

Part 1: Setting Up the Calorimeter

1.1   Set up the soda can calorimeter as shown in the diagram. Insert the straightened out paper clips through the holes in the can to form the support bars, and place it on the iron ring clamped to the ring stand.

1.2   Form the nut support stand by bending half of a paper clip down and inserting into a cork (see Figure 5-1).

1.3   Go on to part 2.

FIGURE 5-1

Part 2: Combustion of Food Product

2.1   Pour 200 mL of water into the soda can. Place thermometer into the drinking hole of the can and secure if necessary.

2.2   Determine the mass of a food product. Record the mass.

2.3   Record the initial temperature of the water in the can.

2.4   Place the food product on the support stand. Ignite the food product using a kitchen match. It may take more than one match to keep it lit.

2.5   When the food product stops burning, stir the water with the thermometer. Record this as the final temperature. Pour out the water from the can. Mass the remaining sample after it cools.

2.6   Repeat the procedure with a new sample. You must use a new batch of water for each trial.

2.7   Repeat the procedure above twice for a total of two experiments.

2.8   Conduct three more trials using a different type nut provided by your TA. Please indicate the type of nut that you are testing on your lab report form.
BE SURE TO WASH YOUR HANDS BEFORE LEAVING THE LAB!!!

Notes

waste put nuts in beaker
of H_2O at front

i mass = 0.408g

i temp. = 23.4°C

final temp = 29.7°C

Δ temp = 6.3°C

final mass = 0.035g

Δ mass = 0.373g

mass of H_2O

200 mL = 200g

density = g/mL

heat produced from nut

$q_{(water)}$ = (specific heat)(quarter)(ΔT)

heat gained from H_2O = heat released from nut

$$q = (1.0 \text{ cal/g°C})(200g)(6.3°C) = 1,260 \text{ cal} = 1.26 \text{ kcal} = 1.26 \text{ Cal}$$

how many cal in one serving

assume 28g in one serving

$$\frac{\text{Cal serving}}{28g} = \frac{1.26}{0.373g} \text{ Cal}$$

Cal serving = 94.58 Cal

i mass = ~~·~~ .542g

i temp = 24.0°C

Final temp = 32.2°C

Δ temp = ~~·~~ 8.2°C

final mass = 0.073g

Δ mass = 0.469g

i mass = .449 g
i temp. = 25.1°C
final temp = 28.5°C

Δ temp = 3.4°C

final mass = .199 g

Δ mass = .250 g

i mass = .947
i temp. 25.7°C
final temp. = 34.7°C

Δ temp. = 9°C

final mass = .232

Δ mass = .715 g

Notes

i mass = .775g

i temp = 26.6 °C

final temp = 37.4 °C

Δ temp = 10.8 °C

final mass = .083g

Δ mass = 0.692g

---

i mass = .741g

i temp. = 243°C

final temp. = 34.7°C

Δ temp = 10.4°C

final mass = .068g

Δ mass = .673g

Laboratory 6
Enthalpy of Neutralization

There are three laws of thermodynamics; the first is that energy is conserved. This law is the basis of today's experiment. A scientific "law" is a theory that has been tested by so many experiments that all scientists agree it is valid. The laws of thermodynamics also have enormous predictive power, allowing scientists to forecast the outcome of many chemical reactions and physical changes.

Along with chemical bonding and molecular structure, energetic relationships provide the most profound organizing principles in chemistry. The conservation of energy also has direct connections with our everyday experience. Note that food packages usually carry a description that includes the calorie content, and heat will raise the temperature of water. Now let's consider the connection that energy from food has to thermodynamics.

HEAT AND TEMPERATURE CHANGE

The relationship between heat, q_p, and temperature change, ΔT, is

(1) $\qquad q_p = C_p \bullet m \bullet \Delta T$,

where C_p is the specific heat capacity for the material (e.g., water) and m is the mass of the sample.[1] For water, it is most accurate to state that

(2) $\qquad C_p = 4.184 \dfrac{J}{°C \bullet g}$,

the definition used in today's experiment.[2] However, most biochemists and biologists still rely on the kilocalorie (1 kcal = 1000 cal) as their primary unit for heat and energy.[3] To a good approximation, one calorie is the unit of heat that is required to raise the temperature of 1.00 gram of water 1.00°C. Thus, the specific heat of water is also

(3) $\qquad C_p = 1.00 \dfrac{cal}{°C \bullet g}$,

[1]The subscript, p, on q_p and C_p is a reminder that the progress of many chemical reactions changes with pressure. The subscript denotes constant pressure.
[2]The conversion factor between joules (J) and calories (cal) is 4.184 J/cal. Joules can be defined with respect to primary standards more accurately than calories, so most chemistry textbooks require that energies be expressed in J than in cal.
[3]The kcal is a convenient yardstick in thinking about biochemical structures and reactions—breaking a typical chemical bond costs about 100 kcal/mol, while the weak interactions that hold an enzyme in a particular three-dimensional shape each contribute about 1–3 kcal/mol toward stabilizing its active form.

Now if we approximate a 150 pound man as a 70 kg bag of water, how much would his temperature increase if he consumes 2,500 kcal of food energy[4] in his daily diet? From eqs. (1) and (3)

$$(4) \quad \Delta T = \frac{q_p}{m \bullet C_p}$$

$$\Delta T = \frac{2{,}500{,}000 \text{ cal}}{70{,}000 \text{ g}} \times \frac{°C \bullet g}{1.00 \text{ cal}} = 36°C.$$

Well, some assumptions must be wrong, or else this poor guy would die of a very high fever! In fact, a lot of the heat would be lost to his surroundings and some of the food energy would be transformed into other molecules in muscle, fat, etc. Still, there is no doubt that a standard diet releases a lot of energy.

Equation (4) also makes it clear that heat and temperature are different, but that a change in heat is manifested in a change in temperature. The conversion between heat and temperature change requires that we know both the mass, m, and the heat capacity, C_p. A small woman, as a less massive bag of water in our 2,500 kcal example, would have a larger ΔT. As a living organism she would have more energy per pound of body weight available to be converted to muscle and fat, or would have to perspire more and do more physical work to dissipate the energy.

TODAY'S EXPERIMENT

Chemical reactions often release or consume heat as the chemical bonds rearrange to form more (or less) stable compounds. Today we will measure the heat released when phosphoric acid reacts with sodium hydroxide,

$$(5) \quad H_3PO_4 + 3 \text{ NaOH} \rightarrow 3 \text{ H}_2O + Na_3PO_4$$

Note that the chemical equation is balanced, with the same number of each type of atom appearing on both sides. The reaction produces heat in an amount that is expressed as the enthalpy,

$$(6) \quad \Delta H° = -167.5 \text{ kJ/mol,}$$

the enthalpy is the heat released by the reaction[5] adjusted for 1 mol of the reactant, H_3PO_4;

$$(7) \quad \Delta H° = q_p/n,$$

where n is the number of moles of the acid that were actually consumed in the reaction. By convention, chemists use negative numbers for reactions that release heat to their surroundings and positive numbers when heat is consumed.

The experiment you will do today is to measure the heat released when you mix 5.0 mL of 0.30 M H_3PO_4 with 5.0 mL of 1.0 M NaOH. Then your partner will repeat the experiment using 10.0 mL of both reactant solutions. You will then compare your result to eq. (6).

[4]Nutrition labels often refer to one Cal (with a capital C) or, confusingly, as one "cal," when they mean one kcal of energy.

[5]The superscript, °, denotes "standard conditions," which here means standard atmospheric pressure = 760 mm of mercury in a barometer, and a solution that starts out with 1.000 M reactant.

Think about three questions before coming to lab, and record in your Notes at the end of the lab module:

<u>Which of the two reactants has the fewest moles of H⁺ or OH⁻?</u> Consider the concentrations and volumes above for the answer. This will limit the amount of heat that will be produced and the resulting temperature change.

<u>When using 10 mL of the reactants, will the heat released be larger, smaller, or the same as for 5 mL?</u> Consider eqs. (1) and (7).

<u>When using 10 mL of the reactants, will the temperature change be larger, smaller, or the same as for 5 mL?</u> Consider eqs. (4) and (7).

PRECAUTIONS

You will be using solutions of sodium hydroxide and phosphoric acid that are diluted about 10-fold in comparison to the concentrated materials. However, they can still cause burns if spilled on your skin or in your eyes. Report any spills to your instructor IMMEDIATELY. Be sure to follow the special directions given by your instructor regarding safety and disposal.

I. Preparation

1.1   *Partner 1* will perform all steps of the experiment with *5.0 mL* of acid and base, and *partner 2* will record the temperature readings in the tables provided at the end of these instructions.

1.2   Each student will attach one original and one copied report sheet to his/her lab report. *The partner mixing the solutions and reading the thermometer* should print his/her name under *"My Name,"* and the *partner recording the data* should print his/her name as *"Partner's Name"* on the report sheet.

1.3   The partners will then switch roles for the steps for the reaction with *10.0 mL* of acid and base.

1.4   Both partners should observe and <u>record answers to the underlined queries in their own Notes</u>. Talk to your instructor if you encounter a problem or have a question.

1.5   Clear off your workspace to remove all books, papers or other items except for these instructions. Partner 2 should turn to the report sheet at the end of these instructions.

1.6   Experiment with the stopwatch to be sure you can use it to measure 30 and 60 second intervals for temperature measurements.

1.7   Turn on the digital thermometer and adjust it to read in °C. Set it on the bench with the tip not touching the bench. After two minutes, <u>record the air temperature reading</u>.

1.8   The digital thermometer does not respond immediately to a change in temperature. At time marked zero, the partner with the warmest hands should hold the tip of the thermometer against the palm of his/her clenched fist. Measure and <u>record the time it takes the thermometer to reach a steady reading</u>.

2. Measuring the temperatures of the solutions

2.1   In the following steps *partner 1* should mix the solutions and read the temperatures. *Partner 2* should record the data in the report sheet's table and keep track of time with the stopwatch.

2.2   Use a squeeze bottle to carefully fill a 15 mL screwtop plastic tube to the 5 mL mark with 0.30 M H_3PO_4; likewise, put 5.0 mL of 1.0 M NaOH in a separate 15 mL screwtop tube. Screw the caps on tightly and label each tube, "Acid" and "Base." Do not hold either tube in your hand for long—you need both of them to start near room temperature. Be careful to avoid spilling either solution. In case of a spill, contact your instructor immediately.

2.3   Put the 15 mL tube containing the base and a separate 40 mL screwtop in holders provided by your instructor. Pour all of the acid into the 40 mL screwtop, which will be your reaction vessel. Put the holders and tubes in convenient places in the center of your workspace. Remove the tops of each tube, but keep them handy. Get a few Kim-Wipes and keep them handy.

2.4   Place the digital thermometer into the 40 mL tube (Acid) and allow the temperature to stabilize for about two minutes.

2.5   You will be making readings every 30 sec at the beginning of the experiment, so read over steps 2.6–2.10 a few times to be sure you know what to do.

2.6   Record the (Acid) temperature at time 0.0 min in the table, start the stopwatch. Quickly remove the thermometer, wipe it with a Kim-Wipe and insert it in the 15 mL tube containing the Base.

2.7   At 0.5 min, record the (Base) temperature. Quickly remove the thermometer, wipe it, and insert it in the Acid tube.

2.8   Repeat steps 2.6–2.7 and record in the appropriate boxes at 1.0 and 1.5 min in the table. Remove the thermometer after the 1.5 min reading.

2.9   At 2.0 min, add all of the NaOH solution to the 40 mL tube, quickly cap, and vortex for 5–10 sec. Open the cap and insert the thermometer.

2.10 Record the temperature at 2.5 min, 3.0 min, then at 1 min intervals until 12 min total time.

2.11 Show the data sheet to your instructor, who will sign it or tell you to repeat the experiment, if necessary.

2.12 Rinse the shaft of the thermometer with water and dry it.

2.13 The partners now switch roles and conduct steps 2.2–2.10 with 10.0 mL of acid and base and a new data sheet.

2.14 When you have finished the measurements with 10 mL and 20 mL total reaction volumes, be sure your instructor has signed both original data sheets.

2.15 Copy the data from your partner's original report table to a new form so that you each have a complete copy of the measurements for both experiments. Attach one original and one copied report to your lab report.

3. Cleanup and Checkin

3.1   Clean up your work area. Dispose of liquid waste and tubes according to your instructor's directions.

3.2   Rinse the shaft of the thermometer with water, and dry it.

3.3   Return the stopwatch and digital thermometer.

3.4   Have the <u>instructor sign and date</u> your Notes.

3.5   Exchange contact information (email, phone) with your partner in case you have questions during the week.

3.6   Do not copy your report from your partner. Collaboration is encouraged, copying calculations or answers is not.

4. Report

- This report is due at the *beginning* of the next regular lab period. Reports turned in after the quiz begins will be marked late. Reports will be marked down 5% for each day that they are late.

- Be sure your name, your TA's name, and your section number are on the front of your report.

- Be sure your partner's name is clearly printed on the page where your report begins.

4.1   Be sure you have written answers to underlined queries in the experiment.

4.2   Plot the two sets of data on the graph attached to each data sheet. See the sample data attached at the end of this experiment.

4.3   Draw a horizontal straight line that best fits the data prior to mixing for each data sheet. Extend the line to 2.0 min. This will give the average temperature of the solutions prior to mixing. Estimate this temperature to the nearest 0.1°C.

4.4   Draw a sloping straight line that best fits the post-mixing data starting with the measurement at 4.0 min and going out to 12.0 min for each data sheet. Extend the line to 2.0 min. This will give the temperature of the solution if you could measure it immediately after mixing and the temperature change was immediate. Estimate this temperature to the nearest 0.1°C.

4.5   Estimate ΔT, the difference between the two temperatures for the 5.0 + 5.0 = 10.0 mL experiment.

4.6   Estimate ΔT, the difference between the two temperatures for the 10.0 + 10.0 = 20.0 mL experiment.

4.7   Calculate the mass of each of the two solutions, using a density of 1.03 g/mL. Density is a conversion factor between volume and mass. Show your work.

4.8   Use eq. (1) to estimate the heat evolved in each solution. Use C_p = 4.184 J/(°C•g). Show your work.

4.9   Calculate the number of moles of the acid and the base in both experiments. Note from eq. (5) that every mole of acid requires 3 moles of base to neutralize it. Will all of the acid react? Explain.

4.10 Use eq. (7) to estimate $\Delta H°$ for each experiment.

4.11 Calculate the average value of $\Delta H°$.

4.12 Calculate the precision of your determination expressed as a % deviation from the average, Av.

$$\%dev = \frac{|Meas1 - Av| + |Meas2 - Av|}{2 \times Av} \times 100,$$

The vertical lines, |…|, mean "absolute values." Ignore any negative signs in calculating the difference from the average.

4.13 Are the slopes of the upper lines different in the graphs for your two determinations? If so, which loses heat faster to the surroundings?

4.14 How would the slopes of the upper lines be different if the reaction tubes were better insulated?

4.15 What would happen to the values determined for q_p and $\Delta H°$ if the concentrations of the acid and base were doubled? Explain.

4.16 Would the precision of the determination of $\Delta H°$ be better or worse if the concentrations were doubled? Explain.

4.17 The accepted value for the enthalpy of neutralizing phosphoric acid is given in eq. (6). What is the % deviation of your determination from this value? Explain why you think the deviation is so large. (By the way, the deviation is large even if you do the experiment perfectly with the apparatus used here.)

5 mL   **Volume Acid**

5 mL   **Volume Base**

Ben Lundgren   **My Name**

Mary Terrinoni   **Partner's Name**

_____ Instructor's signature and date

Time (min)	Temperature (°C) Acid	Temperature (°C) Base
0.0	23.1	
0.5		23.3
1.0	23.1	
1.5		23.3
2.0	Mix	
2.5	27.4	
3.0	27.6	
4.0	27.4	
5.0	27.2	
6.0	27.0	
7.0	26.8	
8.0	26.7	
9.0	26.5	
10.0	26.3	
11.0	26.2	
12.0	26.0	

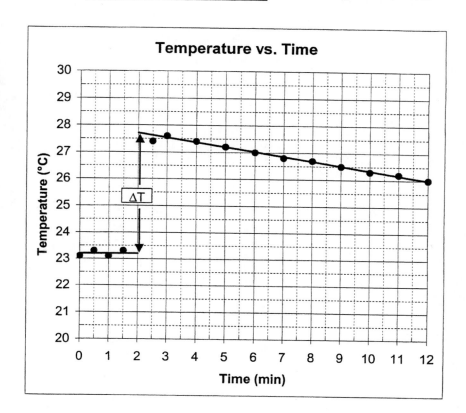

$$\begin{array}{r} 26.2 \\ -24.3 \\ \hline 1.9 \end{array} \qquad \begin{array}{r} 26.8 \\ -24.3 \\ \hline 2.5 \end{array}$$

_____ Volume Acid

_____ Volume Base

_____ My Name

_____ Partner's Name

_____ Instructor's signature and date

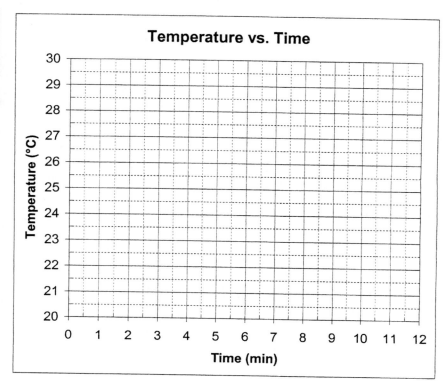

Time (min)	Temperature (°C) Acid	Base
0.0		▨
0.5	▨	
1.0		▨
1.5	▨	
2.0	Mix	
2.5		
3.0		
4.0		
5.0		
6.0		
7.0		
8.0		
9.0		
10.0		
11.0		
12.0		

Notes

4.7 → use total volume

4.8 → $q = nC_p \Delta T$

C_p = heat capacity

4.9 → calculate mols of H^+ and H_3PO_4

4.10 → $\Delta H° = \dfrac{q_p}{n}$ n = # mols H_3PO_4

4.12 → $\dfrac{|(\#-avg)| + |(\#-avg)|}{2\,(avg)} \times 100$

4.17 → accepted value = 167.5 kJ/mol

$\dfrac{accepted - avg}{accepted} \times 100 =$

Notes

Laboratory 7

Issues in Water Quality

We often take the quality of water in our home and work places for granted. Yet there are many issues of water safety and esthetics that are useful to know about. The purpose of this experiment is to perform a test to analyze "hard" and "soft" water. In addition, work on other fundamental issues in water quality will be presented. Some of this will call for students to obtain information from on-line resources.

Most of us have experienced the effect of hard water when we take a bath or shower in a location that gets its water supply from deep wells. Minerals are dissolved as the water passes through rocks and soil in underground rivers, called aquifers. The most common offenders are calcium and magnesium ions, Ca^{2+} and Mg^{2+}, which prevent soap from lathering as much as it does in soft water. Other effects include soap scum left in the tub and a hard scale of $CaCO_3$ in hot water pipes, teakettles, boilers, etc.

Soft water also occurs naturally. One form is in rain water, which originates from surface water that evaporates and leaves the minerals behind. In laboratories it is common to distill or deionize water to remove dissolved salts, but those processes are expensive to implement on a large scale. Home water softeners treat hard water to replace the calcium, magnesium, iron, and other cations with Na^+, which does not precipitate in the presence of soap (to form scum) or heated bicarbonate (to form boiler scale). Many cities get their water from surface sources—lakes or rivers. These sources collect water from underground springs and from rain, which runs off fields, mountains, city streets, etc., picking up minerals and other contaminants along the way. Surface water varies widely in hardness.

If necessary, review the module on **Volume and Mass Measurements** to refresh your memory on the proper use of a micropipet. This is a precision instrument that can be damaged by thoughtless users. Also, refresh your memory about where

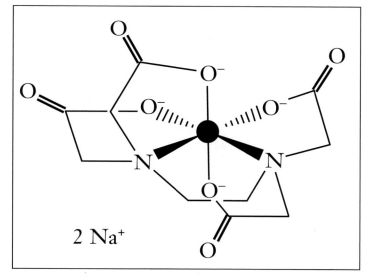

FIGURE 7-1
EDTA complexed with a divalent metal like Ca^{2+} or Mg^{2+} (central black ball). Drawing by J. Zubieta.

the 10 and 50 µL marks are on the pipet tips. Review the sections in your text and lecture notes on molarity and titrations.

You will be using dilute solutions of calcium and magnesium salts and EDTA, which is commonly added to detergents and other household items. Although there is little danger, report any spills to your instructor immediately. Always wear safety goggles and follow your instructor's directions regarding disposal.

Fundamentals of EDTA titrations. You will add a solution containing a known concentration of EDTA to solutions containing (i) magnesium ion, Mg^{2+}, (ii) calcium ion, Ca^{2+}, and (iii) a mixture of the two metal ions. EDTA is an abbreviation for ethylenediaminetetraacetic acid. It has four acidic protons that can be replaced by counterions in salt forms of the molecule. The salt form that applies in today's experiment is Na_4EDTA, where all of the acidic protons have been replaced by sodium ions. The anion, $EDTA^{4-}$, forms high-affinity structures (called chelates, key'-lātes) that wrap around many metal ions, forming six bonds with Ca^{2+} or Mg^{2+}. A schematic view of the structure is shown in Figure 7-1. The important transformations are represented in:

(1) $Ca^{2+} + EDTA^{4-} \rightarrow [Ca \cdot EDTA]^{2-}$

(2) $Mg^{2+} + EDTA^{4-} \rightarrow [Mg \cdot EDTA]^{2-}$

(all of these species are colorless)

The affinity of $EDTA^{4-}$ for Ca^{2+} is higher than for Mg^{2+}. When a mixture of Ca^{2+} and Mg^{2+} is titrated with $EDTA^{4-}$, $[Ca \cdot EDTA]^{2-}$ forms first, then $[Mg \cdot EDTA]^{2-}$ begins to form as more $EDTA^{4-}$ is added. At the endpoint of the titration all of the Mg^{2+} is consumed, as well.

To distinguish the endpoint of the titration, we add the organic dye indicator, Eriochrome Black T (EBT). The dye forms the blue-colored $HEBT^{2-}$ anion under the conditions used in this experiment.[1] EBT^{3-} forms a tighter complex with Mg^{2+} than Ca^{2+}, but each is bound more weakly than the EDTA complexes.

(3) $Mg^{2+} + HEBT^{2-} \rightarrow [Mg \cdot EBT]^- + H^+$

(4) $Ca^{2+} + HEBT^{2-} \rightarrow [Ca \cdot EBT]^- + H^+$

(blue)   (red)

The metal-EBT complexes each have a red color. Thus, ***the titration proceeds from a red-wine color*** (early in the titration only a little EDTA has been added, so there is a lot of free Ca^{2+} and Mg^{2+} to bind EBT^{3-}), ***to purple just before the endpoint*** (where the red color of $[Mg \cdot EBT]^-$ mixes with the blue of free $HEBT^{2-}$), ***and finally to a permanent blue*** (when the last of the $[Mg \cdot EBT]^-$ has been converted to $[Mg \cdot EDTA]^{2-}$; of course, Ca^{2+} is already present as $[Ca \cdot EDTA]^{2-}$ because it has a higher affinity for EDTA than Mg^{2+}).

CALCULATIONS FOR WATER HARDNESS. EDTA titrations are similar to the acid-base titrations we performed in an earlier module. Recalling the handy formula:

(5) $M_1V_1 = M_2V_2,$

let solution one be EDTA, with a value for $M_1 = 0.0100$ M. V_2 is the volume of $Mg^{2+} + Ca^{2+}$ solution, and V_1 will be known at the end of the titration. Then you must solve for

---

[1] The solution has been adjusted to be mildly basic (pH 10), using a "buffer" composed of NH_3 and NH_4Cl in water.

$M_2 = [Mg^{2+}] + [Ca^{2+}]$. Note that the individual concentration of either metal ion can be zero—the method presented here measures only the "total hardness" of the water. Other procedures would be necessary to determine individual values for $[Mg^{2+}]$ and $[Ca^{2+}]$.

I. PREPARATION

1.1 Label four snap top tubes as "EDTA," "Mg," "Ca," and "Mix;" the latter solution has a mixture of the two ions. Obtain about 1.5 mL of EDTA and about 1 mL of the other solutions.

1.2 Label a snap top as "EBT," and obtain 100 μL.

1.3 Partners should label six snap tops, "1," through "6." Also label the tubes with the first letter of your first name. Partner 1 will perform all steps of the experiment on the odd-numbered tubes, and partner 2 on the even numbers. Both partners should observe; record answers to the underlined queries on your own Notes, including the measurements made by your partner. Talk to your instructor if you encounter a problem or have a question.

1.4 Snap the tops shut for any mixing steps. You cannot lose any of the solutions for the titrations to give accurate results.

1.5 When making additions of EDTA, *do not touch the solution being titrated with the pipet tip*, or you will have to use a new tip for the next addition.

1.6 When drawing liquid into the micropipet, look at the tip to be sure there are no air bubbles and to verify that the tip is filled to about the right level.

1.7 When setting accurate volumes, eliminate "backlash" errors. Turn the top knob to a larger volume than desired (by about 1/3 of a turn), then dial slowly down and stop when the desired value is displayed in the volume window.

1.8 Unlock the wings of the micropipet, set it for 150.0 μL, and verify that the rate controller is in the middle of its range.

2. COLOR CHECKS

2.1 Put 150 μL of the Mg^{2+} solution in tubes 1 and 2 and do the following steps.

2.2 Set the rate controller on the pipettor to very slow (–) and allow about 10 seconds for the tip to fill to the 10 μL mark. Do this each time you pipet the viscous EBT solution. Reset the rate controller to mid-range to pipet the other solutions. Add 10 μL of the EBT solution to each tube, close the cap, and mix. What is the color? What ion gives rise to the color? This will be the color at the beginning of any titration containing Mg^{2+} or Ca^{2+}. Keep tube 1 as a reference for comparison to distinguish the color early in titrations.

2.3 Add 200 μL of the EDTA solution to tube 2, close the cap, and mix. What is the color? What ion gives rise to the color? This will be the color at the end of any titration containing Mg^{2+} or Ca^{2+}. Keep tube 2 to compare with other titrations after the endpoint. Whether the *color* is blue, red, or purple is the key to distinguishing the endpoint. The color will be more intense if the concentration is higher, or less intense if the concentration is lower, but the color blue is always blue and red is always red.

2.4 Put 150 μL of the Ca^{2+} solution in tube 3 and do the following steps.

2.5 Add 10 μL of the EBT solution, close the cap, and mix. What is the color? What ion gives rise to the color? Is the *color* the same as in step 2.2?

2.6 Add 200 μL of the EDTA solution, close the cap, and mix. What is the color? What ion gives rise to the color?

2.7 Note that the color change in this experiment is more difficult to distinguish than in acid-base titrations using phenolphthalein as the indicator. There may be a fairly broad range over which the color changes from red to purple to blue. The appearance of a permanent blue signifies that the endpoint has been reached.

3. DETERMINING TOTAL HARDNESS = $[Ca^{2+}]$ + $[Mg^{2+}]$

3.1 Tube 4 will be titrated with rather large additions of EDTA. This will give you a good idea of where the *endpoint* will occur. Then for tubes 5 and 6 you can use large additions until you are close to the endpoint and 2.0 µL additions in the vicinity of the endpoint.

3.2 Dispense 150.0 µL of the mixed solution into tubes 4, 5, and 6.

3.3 Use a new pipet tip to add 10 µL of EBT solution to each tube. Record <u>what happens</u>. Briefly mix the contents with the vortex mixer.

3.4 Use a new pipet tip and add 100.0 µL of EDTA to tube 4. Record <u>what happens</u>. Briefly mix the contents with the vortex mixer.

3.5 Add 10.0 µL of EDTA. Record <u>what happens</u>. Briefly mix the contents with the vortex mixer. Wait about a minute to judge the final color against the saved tubes 2 and 3.

3.6 Repeat step 3.5 until a permanent color change occurs. Record the <u>total volume of EDTA added just prior to the permanent color change</u>.

3.7 With tube 5, add the total amount of EDTA from 3.6. <u>Record in your Notes which partner did the work with tube 5</u>.

3.8 *Carefully* reset the pipet for 2.0 µL—you can break it if you try to set it below 0.0. Practice pipetting 2.0 µL of water until you are confident that you can do it with the EDTA solution—check to be sure a small amount of liquid goes into the tip.

3.9 Now add 2.0 µL increments of EDTA to tube 5, mix and watch for a permanent color change. Your partner should <u>make a note in their Notes</u> after each increment to be sure you don't get confused.

3.10 Record the <u>total volume of EDTA added just prior to and just after the permanent color change</u> to blue. Take the average of these volumes as the equivalence point.

3.11 *CAREFULLY* reset the pipet to a larger volume. As you change the top knob, **watch the volume indicator.** Never go below 0.0 µL. If you are unsure about this, see your instructor. Reset the pipet to the total volume found in 3.6.

3.12 With tube 6, repeat 3.7 to 3.11. <u>Record in your Notes the other partner's name who did the work with tube 6</u>.

3.13 Do the volumes determined for tubes 5 and 6 differ from the each other by more than 20%? If so, repeat steps 3.1 to 3.11, obtaining a fresh 1 mL of EDTA solution.

4. CLEANUP AND CHECK IN

When you have finished all parts of the module:

4.1 Clean up your work area. Dispose of liquid waste, tips, and tubes according to your instructor's directions.

4.2 Return the pipettor and other supplies to your instructor.

4.3 *Clearly print* your <u>lab partner's name</u> and the experiment name in your Notes.

4.4 Have your <u>instructor sign and date</u> your Notes.

REPORT:

5. CALCULATIONS FOR TOTAL HARDNESS

5.1 Use eq. (5) to calculate the average value for $M_2 = [Mg^{2+}] + [Ca^{2+}]$ for tubes 5 and 6. Record the average M_2 in your Notes. Also calculate and record the average percent deviation for your two determinations. See the module on acid-base titrations if you have forgotten how to calculate average percent deviation.

5.2 It is not common practice for water quality managers to report concentrations of dissolved materials in units of molarity. Rather, amounts are usually expressed in parts per million (ppm; 1 mg/L = 1 ppm) or parts per billion (1 μg/L = 1 ppb). Assume that all of the hardness in M_2 comes from Ca^{2+}, calculate the amount of Ca^{2+} in ppm, and record in your Notes and Report sheet.

5.3 Water hardness is usually set in five categories (see table). Into which category would today's sample fall?

Classification	ppm
Soft	0-17.1
Slightly hard	17.1-60
Moderately hard	60-120
Hard	120-180
Very Hard	over 180

6. OTHER CONTAMINANTS IN FRESH WATER

As resource material, refer to the following Web sites:

http://www.ocwa.org

http://www.upstatefreshwater.org/html/onondaga_lake.html—note that the last part of this address has an underbar: onondaga_lake.

6.1 What were the average levels of Ca^{2+} and Mg^{2+} found in Otisco lake in the most recently reported year? To what classification does this correspond in the table provided?

6.2 Cryptosporidium and Giardia are sometimes present in lakes, rivers, and streams. Unfortunately, they are not killed by chlorination. Why is that a problem?

6.3 Roughly what fraction of the water input to Onondaga lake comes from the Syracuse metropolitan sewage district?

6.4 What pollutant is the cause of a ban on eating fish caught in Onondaga lake?

CRUD

Lyle

EBT = indicator

Notes

[EDTA] = 0.0100M

5.1 a) $M_1V_1 = M_2V_2$

5.1 b) $$\frac{|\#5-ave| + |\#6-ave|}{2\ (Ave)}$$

5.2) 1ppm = 1mg/L

6.1-6.9   Find online

Before eq. point   Red.

After eq. point   Blue.

2.5) Purple, no
2.6) Blue, Ca^{2+}

2.2 a) red
  b) Mg^{2+}
2.6) a) blue
  b) Ca^{2+}

3.3) each tube turns pink

3.4) Still pink. Not much change

3.5) No change

3.6) 150μL

3.7) Everybody

3.10) 150μL prior eq.
       152μL after eq.

3.12) Everybody

3.13) Equivalence, same, as 3.10

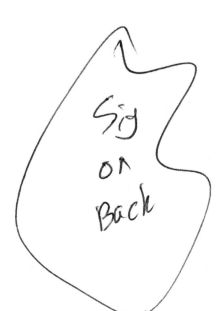

Sig
on
Back

Cleared by
M Reyes
10/23

Notes

Laboratory 8

Atomic and Molecular Structure, Part 1

Bonding and molecular structure provide the entry to a sophisticated understanding of chemistry. Today's lab will reinforce concepts presented in your lecture text about the properties of atoms and *Lewis structures*. Review those chapters before coming to lab. Today's lab also introduces a computer program to build and visualize chemical compounds. The program will be used in the next lab module to study the details of molecular structures. If you have not had a previous chemistry course it is a good idea to read through this document twice and see your TA or instructor to answer questions you may have. There will be a quiz near the end of the lab period on the material presented here.

The three-dimensional arrangement of atoms in molecules is central to the chemical properties of compounds. That arrangement is determined by the way electrons are distributed between the nuclei of the atoms involved in covalent and ionic bonds. Chemical reactions involve rearrangements in the distribution of electrons.

This and the next module will use a molecular modeling program, **ChemSketch,** from ACD Labs.[1] This module will focus on the valence electron configurations of atoms and Lewis electron dot structures for compounds, and use the program to draw simple structures. The next module will progress to more complicated molecules.

If you have a laptop computer, please come to the laboratory meeting with the program already loaded. If you don't have a laptop, you will work with a partner who does have one. You may also want to load the program on another computer—it may be helpful in studying for your lecture exams and in future chemistry courses.

BEFORE CLASS BEGINS

The program can only be downloaded on campus.[2] It is suggested that you temporarily save the file on your desktop. Once downloaded, it can also be installed on a home computer away from campus. Follow the instructions to do the installation. After installation is complete, copy the **ChemSketch** shortcut to the desktop. There will be a short quiz part way through today's module.

[1]http://www.acdlabs.com/ Advanced Chemistry Development, Inc.
[2]See the course Web site for the download.

THE ELECTRONIC STRUCTURE OF ATOMS

The *periodic table* in Figure 8-1 looks a little different from the one inside the cover of this manual. This version emphasizes the electronic arrangement of the atoms. The atomic number at the top of each entry gives the total number of electrons for an atom. The atoms within each vertical column tend to react similarly to each other because they have identical outer electron configurations. It is these outer or *valence* electrons that participate in most chemical bonds. For instance, the *halogens* all have seven outer shell electrons (elements 9-fluorine, 17-chlorine, 35-bromine, 53-iodine, and 85-astatine). The eight columns of elements in the large top block in the figure are called *representative elements* or *main group elements*. Studying their electron configurations and reactions goes a long way toward understanding all chemical reactions. Only eleven elements (noted in large italic type) are highly abundant in living systems, and they are all representative elements. The other thirteen elements in italics are also common in living organisms. The elements shown in italics and those in letters that are larger than their neighbors are often discussed in introductory chemistry courses.

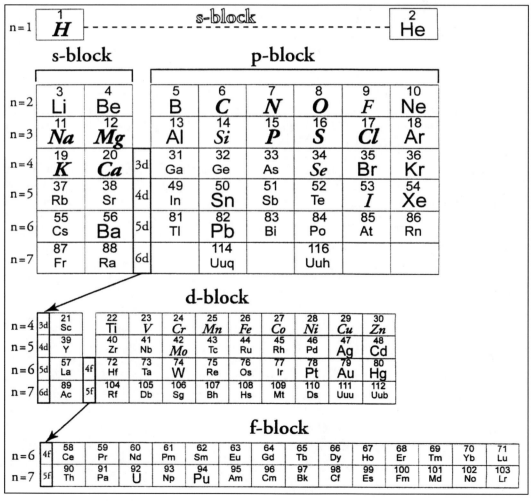

FIGURE 8-1
The chemical elements arranged by their valence electrons.

The arrangement of electrons in atoms follow the rules of quantum mechanics, which stipulate that each electron in an atom must have a unique combination of the four quantum numbers, n, l, m_l, and m_s (see your textbook for definitions of the quantum numbers). The electrons in each horizontal row of the chart have the same principal *quantum number, n,* in the valence *shell.* The value of n for this shell is given at the left of the rows in Figure 8-1. The number of valence electrons increases left to right in each row. The *atomic orbital* definitions, s, p, d, f, correspond to increasing values of the second *quantum number, l;* each orbital can hold two electrons. There is one *s-orbital* ($l = 0$) for every shell of electrons. Hydrogen and helium are special in that they have only s-electrons. For all the other shells, there are three *p-orbitals* ($l = 1$). The representative elements all have from one to eight electrons in their outer shells. Atoms are especially stable when their outer shells are filled. Thus, the *noble gas* elements are nearly unreactive (elements 2-helium, 10-neon, 18-argon, 36-krypton, 54-xenon, and 86-radon).

ELECTRON CONFIGURATIONS

The shorthand for the electronic configurations of atoms denotes the occupancy of shells and orbitals. The examples in Figure 8-2 make it clear how to do this. *Lithium* has three electrons—the first shell is filled ($1s^2$) and its second shell has only one electron ($2s^1$), so its full configuration is, $1s^2 2s^1$. Fluorine has nine electrons; four of them fill the 1s and 2s orbitals ($1s^2 2s^2$), and five occupy the 2p *subshell* ($2p^5$), so its full configuration is $1s^2 2s^2 2p^5$. Sodium has 11 electrons, distributed as $1s^2 2s^2 2p^6 3s^1$. This shorthand notation gets more and more complex for atoms with a lot of electrons, so it is abbreviated further by noting the noble gas core and specifying the details only of the outer, reactive, electrons. Thus, lithium is $[He]2s^1$, sodium $[Ne]3s^1$, chlorine $[Ne]3s^2 3p^5$, etc. <u>Fill in the blank boxes in Figure 8-2. and later copy them to your lab report.</u>

In comparing atoms with increasing numbers of electrons, the s-block fills first, then the p-block fills for shells 2 and 3. For n = 3 to 7, five *d-orbitals* ($l = 2$) are available but, as is seen in Figure 8-1, the 3d subshell is not occupied until the 4s subshell is filled. A similar gap occurs between 5s and 5p, 6s and 6p, and 7s and 7p. This *d-block* in Figure 8-1 contains the *transition elements.* Note that the first of these, Sc, has 21 electrons and follows Ca, element 20, a

Total Electrons	1							2
Element	H							He
Valence electrons	1							2
Full configuration	$1s^1$							$1s^2$

Total Electrons	3	4	5	6	7	8	9	10
Element	Li	Be	B	C	N	O	F	Ne
Valence electrons	1		3		5		7	8
Full configuration	$1s^2 2s^1$		$1s^2 2s^2 2p^1$		$1s^2 2s^2 2p^3$		$1s^2 2s^2 2p^5$	$1s^2 2s^2 2p^6$
[Core] + valence	$[He]2s^1$		$[He]2s^2 2p^1$		$[He]2s^2 2p^3$		$[He]2s^2 2p^5$	$[He]2s^2 2p^6$

Total Electrons	11	12	13	14	15	16	17	18
Element	Na	Mg	Al	Si	P	S	Cl	Ar
Valence electrons	1			4			7	8
[Core] + valence	$[Ne]3s^1$			$[Ne]3s^2 3p^2$			$[Ne]3s^2 3p^5$	$[Ne]3s^2 3p^6$

Figure 8-2
Electron configurations for the first three rows of representative elements.

representative element, in the top block. The electron configuration for Ca is [Ar]$4s^2$ and for Sc is [Ar]$3d^1 4s^2$. The chemical differences between the d-block elements are more subtle than for the representative elements. That is because they all have an outer shell containing only s-electrons. They are all *metals*, but differ from each other in interesting and often useful ways. Element, 31-Ga, which comes right after the first d-block insertion in Figure 8-1, has the configuration [Ar]$3d^{10}4s^2 4p^1$ and Br is [Ar]$3d^{10}4s^2 4p^5$.

In the standard periodic chart inside the cover of this manual, the transition elements are inserted in one long row, continuous with their s- and p-block neighbors. The d-block is displaced in Figure 8-1 to emphasize the close relationship of the s- and p-block elements, and to show the similarity between the d-block and the f-block, or inner transition elements. For n = 4 to 7, there are seven f-orbitals (*l* = 3). They don't begin to be occupied until after 57-La and 89-Ac (see Figure 8-1). Thus, element 72, Hf, has the configuration, [Xe]$4f^{14}5d^2 6s^2$, while element 81, Tl is [Xe]$4f^{14}5d^{10}6s^2 6p^1$. Don't worry about writing complex electron configurations similar to these last two. Focus instead on the way the d-block and f-block insertions are made in Figure 8-1.

ORBITAL OCCUPATION

As was mentioned earlier, four quantum numbers, n, *l*, m_l and m_s, define the characteristics of electrons in atoms. No two electrons in an atom can have the same values for all four of the quantum numbers. So how are the electrons within an orbital different from each other? We have discussed n and *l* so far, so we know that each row of the periodic chart has its own value for n, and that *l* has a different value for each class of orbitals (s, p, d, f). The s-orbitals all have the same value for m_l, but each of the three p-orbitals has a different m_l. The five d-orbitals all have different m_l values from each other, as do the seven f-orbitals. For now, we will discuss only the s- and p-orbitals of the representative elements in the top block of Figure 8-1. Each of the p-orbitals has directional qualities that will be important in the next lab module. For the time being, just note that each of the p-orbitals confine electrons to different regions of space. The final quantum number, m_s, defines the *spin* of an electron. It can have only two values, +½ (spin up, ↑), or -½ (spin down, ↓). Now let's apply this to the first three rows of representative elements.

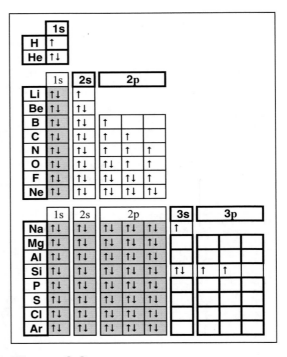

Figure 8-3
Orbital occupancy for the first three rows of representative elements. Closed shells are in gray.

Figure 8-3 shows orbital occupancy diagrams for the same set of atoms as in Figure 8-2. Starting at the top, hydrogen has a single electron and helium has two *paired* electrons (one spin up, one spin down).[3] That fills the 1s orbital, which is also filled in the next set of elements. This *closed shell* of electrons does little to contribute to the chemistry of these atoms. The valence shell consists of 2s and 2p orbitals for Li to Ne. Electrons are charged, so whenever possible they occupy regions of space that place them as far apart as possible. That is why the 2p orbitals of nitrogen each contain one electron.[4] For the bottom block in Figure 8-3, the first and second shells are closed. Fill in the missing arrows in the 3s and 3p levels of Figure 8-3, and later copy them to your lab report.

Take another look at Figures 8-2 and 8-3, and notice that these are two alternatives that represent the same principles. Closed shells in Figure 8-3 correspond to noble gas cores in Figure 8-2, and the emphasis on valence electrons is clear in both. The atomic configurations of Figure 8-2 have the advantage of compactness. However, the presentation in Figure 8-3 is visually appealing and will be useful in understanding the directional properties of chemical bonds.

ELECTRONEGATIVITY

The *octet* rule states that atoms tend to react so they end up being associated with eight valence electrons, a filled outer shell.[5] Atoms that tend to pick up extra electron(s) are highly *electronegative*, while those that easily surrender outer electron(s) are least electronegative. Figure 8-3 shows that if fluorine (F) had one more electron, its electron configuration would be the same as neon (Ne). This is true for all of the halogens—look at the orbitals you filled in for Cl and Ar in Figure 8-3. Thus, reactions are favored if they convert F → F⁻ and Cl → Cl⁻, with the halogen atoms picking up an extra electron. On the other side of the periodic table are atoms like Na and K. If Na lost its lone outer shell electron, it would have the eight electron configuration of Ne. Thus, reactions are favored if they convert Na → Na⁺ and Li → Li⁺. The reaction where sodium and chlorine react to produce the *ionic compound*, NaCl, is very strongly favored. Sodium wants to give up an electron and chlorine wants it—party time! The arrangement of atoms in a crystal of NaCl is easily explained as the Na⁺ and Cl⁻ ions are held close to each other by the attraction of positive and negative charges.

Many atoms have little tendency to form ionic compounds; they have in-between values for electronegativity. Carbon, halfway between Li and F in the periodic table, is the prototype. It would have to gain four electrons to achieve an outer shell like Ne or to lose four to be like He. That's

---

[3]According to *Hund's Rule* each p-orbital is occupied by a single electron and each has the same spin until there are more than three electrons. Then electrons must pair within an orbital. Look for this trend in Figure 8-3 for C, N, then O, F, and Ne.

[4]It is only for convenience that the p-orbitals are shown as filling from left to right—they all have the same energy. The 2s orbitals have lower energy than 2p so two electrons occupy 2s before any occupy the 2p. The same is true for 3s and 3p.

[5]Hydrogen and helium are special cases, where the octet rule becomes a "duet" rule—two electrons, rather than eight, fill their outer shell. There are other exceptions to the octet rule, as well.

a big price to pay in either direction, so C tends to share electrons and make covalent bonds. Most atoms will share electrons with others to make stable covalent bonds.

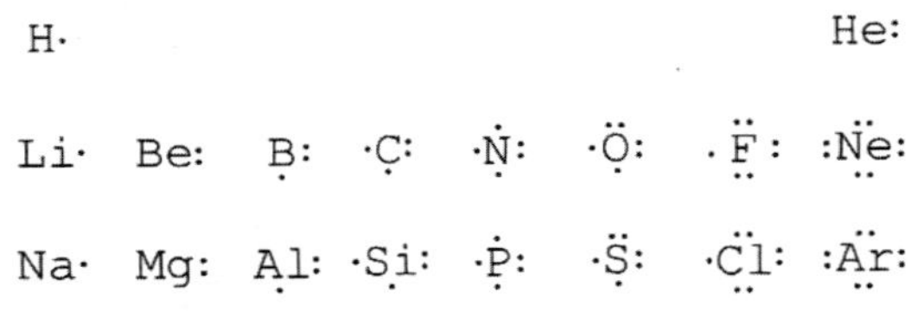

H·   He:

Li·   Be: B: ·C: ·N: ·Ö: ·F: :Ne:

Na·   Mg: Al: ·Si: ·P: ·S: ·Cl: :Ar:

FIGURE 8-4
Electron-dot pictures for the first 18 atoms.

LEWIS ELECTRON-DOT STRUCTURES

The electron-dot representations of atoms in Figure 8-4 closely resemble the orbital occupation diagrams in Figure 8-3. The electron-dot picture allows a quick look at what electrons are available to participate in ionic and covalent bonds. Electron-dot structures of compounds offer a good introduction to chemical bonding. A more accurate view will be provided in the next lab module.

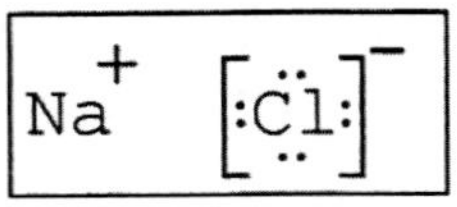

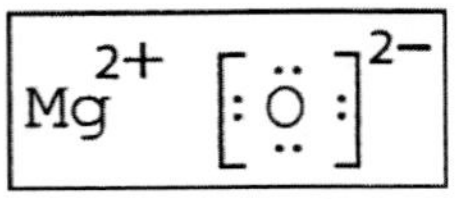

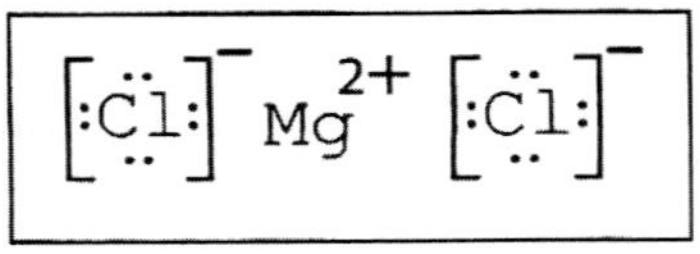

Ionic bonding. The electron dot structures for ionic compounds are similar to the diagram at the right. A neutral chlorine atom has picked up an electron from a sodium atom to produce a chloride anion with eight outer electrons, and a sodium cation with an empty outer shell and [Ne] core. Magnesium oxide provides a similar example. Looking at Figure 8-4, it can be seen that Mg can donate two electrons to oxygen to generate the doubly charged ions. Still another example is magnesium chloride, where each chlorine takes an electron from magnesium. <u>In your Notes, write electron dot formulas for the ionic compounds, LiF and Na₂S.</u>

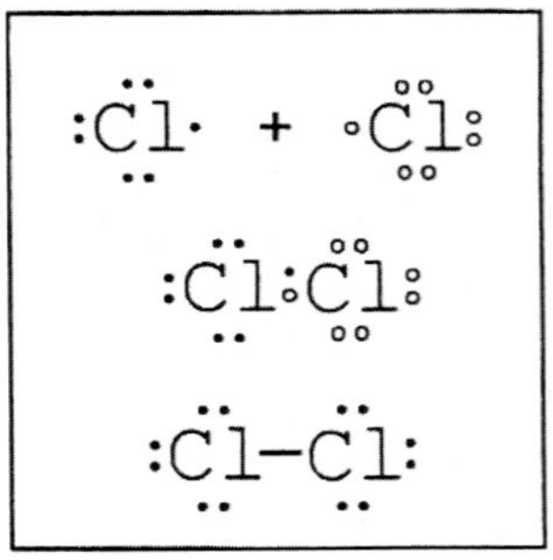

$$Na^+ \quad \left[:\ddot{C}l: \right]^-$$

$$Mg^{2+} \quad \left[:\ddot{O}: \right]^{2-}$$

$$\left[:\ddot{C}l: \right]^- Mg^{2+} \left[:\ddot{C}l: \right]^-$$

Covalent bonding. The Lewis formulas for covalent compounds are more interesting. As a first example, consider dichlorine—the compound often used to kill bacteria in drinking water. The top picture shows two chlorine atoms, one with dots and the other with circles to represent their seven valence electrons. They each need one more electron according to the octet rule. The middle panel shows that each can be associated with eight electrons if they contribute one electron to a shared pair. That constitutes a *single bond,* indicated by a single line in the bottom panel. The electrons denoted by the dash are called a *bonding pair,* while each of the other paired electrons are called *lone pairs.* Of course, electrons are not distinguishable from each other, so it is okay to picture them as all dots. However, using dots, circles, and x's for the electrons from different atoms helps to ensure you have counted the total number of valence electrons correctly.

$$:\ddot{C}l· \; + \; ·\overset{\circ}{\underset{\circ}{C}}l\overset{\circ}{\underset{\circ}{:}}$$

$$:\ddot{C}l \overset{\circ}{\underset{\circ}{:}} \overset{\circ}{\underset{\circ}{C}}l\overset{\circ}{\underset{\circ}{:}}$$

$$:\ddot{C}l-\ddot{C}l:$$

Writing electron dot structures is like solving a puzzle, with a few rules to help make the pieces fit. One is to be sure the total number of electrons is accounted for in the final structure. Another is that complex compounds are likely to have C, O, N, or S as central atoms. Your text gives some other rules, which you should review, but mostly it takes practice.

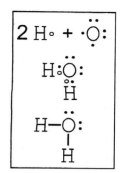

Water is a good example with oxygen as a central atom, and hydrogen (which can only share one pair of electrons). The top picture shows that the atoms start with a total of eight valence electrons. The center panel shows that they can all be accounted for with two pairs of bonding electrons. There are also two lone pairs on the oxygen atom; these turn out to be very important in the way that water interacts with biomolecules and other compounds.

Double and triple bonds are also common in covalent compounds. The next examples illustrate them with dioxygen with two bonding pairs and dinitrogen with three. These are the most abundant components of the air we breathe. Notice also that we are free to rearrange the places where we write the lone pair electrons. Their positions in the diagrams shown here are close to the spatial positions where they are localized in the real molecules. In your Notes, write Lewis structures for CH_4 and CO_2.

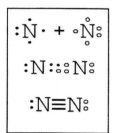

The next examples are ammonia (NH_3) and the ammonium ion (NH_4^+). The three shared pairs in ammonia are apparent, with nitrogen bringing five valence electrons and the hydrogens contributing a total of three. Ammonium ion has the same number of valence electrons because of the positive overall charge on the ion. Notice that ammonia's lone pair provides the bonding electrons for the fourth hydrogen.

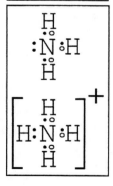

Your instructor will go through a number of examples of other Lewis electron dot structures. This will be followed by a quiz.

INTRODUCING THE PROGRAM

Your instructor will go through the following material in class. Double click the **ChemSketch** icon to open the program. Size the window to fill the screen. Then activate the **3D Viewer**. Navigation buttons at the bottom of both programs allow you to jump between the two programs. Click the one that says, **ChemSk,** to return to the drawing program. Spend a few minutes identifying these icons on the **ChemSk** toolbars:

- **Undo**
- **Select/Move**
- **Select/Rotate/Resize**
- **3D Rotation**
- **Draw Normal**
- **3D Optimization**
- **Periodic Table of Elements**

Do the same with these icons on the *3D* toolbars:

- **3D Rotate**
- **Rotate**
- **Move**
- **Resize**
- **Wireframe**
- **Stick**
- **Ball and Stick**
- **Spacefill**
- **with Dots**
- **Bond Length**
- **Angle**
- **Invert Center**
- **Auto Rotate**
- **Auto Rotate and Change Style**

Return to **ChemSk.** Add to the list of available atoms that can be used in your molecular drawings by clicking the **Periodic Table** icon. On the right side near the bottom is an icon that looks like a pushpin tack **(Change Navigation Mode)**. Click it a couple of times and leave it in the "pulled out" mode. Add these to the list of atoms: **F, Si;** verify that the new atoms appear along the left side of the screen, along with **C, H,** etc. Remove anything you may have accidentally put on the drawing board by selecting all objects **(Ctrl A)**, then hit the **Delete** key.

MEASURING BOND LENGTHS AND BOND ANGLES (PART I).

Halogen single bonds. Be sure **Structure** and **Draw Normal** are activated.

1.1 Click on the atom **Br.** In the drawing space click and hold the left mouse button and drag to the right. You will see dibromine displayed. Then click **3D Optimization** three times to let the program optimize the bond length.[6] Click **Copy to 3D** and display as **Ball and Stick.** Click **Distance,** one of the bromine atoms and then the other. <u>Write the Br-Br bond distance in your Notes to the nearest 0.1 Å.</u>[7]

1.2 Back in **ChemSk,** click on the atom, **Cl,** then each of the Br atoms in the molecule, rerun the optimizer (twice), and measure the Cl–Cl bond distance in the **Viewer;** <u>write the bond distance for dichlorine in your Notes to the nearest 0.1 Å.</u>

1.3 Create and optimize difluorine, measure and <u>record the bond distance in your Notes.</u> <u>Explain the trend in bond lengths for the halogens.</u> Include in your explanation the charges on the nuclei and the number and shell occupancy of core electrons.

[6]The optimizer algorithm changes bond lengths and angles to seek the lowest energy configuration of the molecule. Bond distances are set within about 0.1 Å of their true values.

[7]Most chemists, biochemists, and biologists express interatomic distances in Angstrom units because typical bond lengths are about 1 Å. Many textbook publishers insist on units of nanometers or picometers; 1 Å = 100 pm = 0.1 nm.

Double and triple bonds.

1.4 Now draw dioxygen. The program assumes that there will be a single bond between the oxygen atoms and puts hydrogens at each end. Now click on the central bond to change it to a double bond. Click it again and it goes to a triple bond, then a third click brings it back to a single bond. Click it again to double, optimize (three times), and measure the bond distance. <u>Record the value in your Notes</u>.

1.5 Do the same thing with dinitrogen, creating a triple bond, and <u>record the bond distance</u>. In **ChemSk,** hit **Ctrl A** and **Delete.**

The next module will continue to use the program to draw more complex molecules.

Notes

CH_4

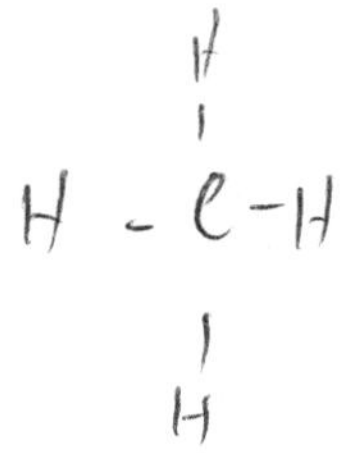

$$H - \underset{\underset{H}{|}}{\overset{\overset{H}{|}}{C}} - H$$

$CO_2 = 4 + 2(6) = 16$

$$O - C - O$$

C_2H_6

$2(4) + 6 = 14$

Notes

Laboratory 9

Atomic and Molecular Structure, Part 2

This module highlights principles of covalent bonding using a molecular modeling program, **ChemSketch,** from ACD Labs.[1] The previous module discussed the valence electron configurations of atoms and Lewis electron dot structures, and used the program to measure bond distances for some diatomic compounds. This module progresses to more complicated molecules.

If you have a laptop computer, please come to the laboratory meeting with the program already loaded. If you don't have a laptop, you will work with a partner who does have one. You may also want to load the program on another computer—it may be helpful in studying for your lecture exams and in future chemistry courses. The program can only be downloaded on campus.

It is recommended that you practice using the program prior to coming to this lab session. Sections 1–3 are helpful in learning the basics.

INTRODUCING THE PROGRAM

Your instructor will review aspects of the menus in class. Double click the **ChemSketch** icon to open the program. Size the window to fill the screen. Then activate the **3D Viewer**. Navigation buttons at the bottom of both programs allow you to jump between the two programs. Click the one that says, **ChemSk,** to return to the drawing program. Spend a few minutes identifying these icons on the **ChemSk** toolbars:

- **Undo**
- **Select/Move**
- **Select/Rotate/Resize**
- **3D Rotation**
- **Draw Normal**
- **3D Optimization**
- **Periodic Table of Elements**

[1]http://www.acdlabs.com/ Advanced Chemistry Development, Inc.
[2]See the course Web site for the download, and the previous module for instructions on installing the program.

Do the same with these icons on the **3D** toolbars:

- **3D Rotate**
- **Rotate**
- **Move**
- **Resize**
- **Wireframe**
- **Stick**
- **Ball and Stick**
- **Spacefill**
- **with Dots**
- **Distance**
- **Angle**
- **Invert Center**
- **Auto Rotate**
- **Auto Rotate and Change Style**

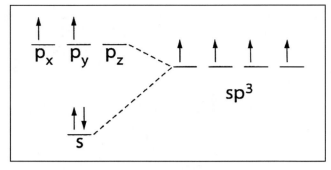

FIGURE 9-1
The ground-state orbital configuration of carbon's valence electrons is shown on the left,[3] and sp^3 hybrid orbitals on the right. Both diagrams satisfy the Pauli principle.[4] Energy increases along the vertical axis.

I. BOND LENGTHS AND BOND ANGLES

The "octet rule," a theory that was discussed in the previous module can be expressed in Lewis electron dot formulas for compounds. This theory can tell us about the likely atomic composition of most chemical compounds, but the theory makes incorrect predictions or is ambiguous for certain compounds. It also tells us little about the lengths of chemical bonds or the three-dimensional shapes of molecules.

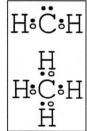

FIGURE 9-2

Valence bond theory makes better predictions of the shapes of molecules than the octet rule. Last time we discussed the rules for how valence electrons occupy atomic orbitals. This is illustrated at the left of Figure 9-1 for carbon. This part of the diagram suggests that C ought to form CH_2 according to the Lewis formula at the top of Figure 9-2. However, carbons almost always make four bonds, as in the Lewis formula for methane (Fig. 9-2, bottom). The bonding is more consistent with the electron occupancy diagram shown at the right of Figure 9-1, where a mixture of one s- and three p-orbitals are considered to create four *sp^3 hybrid orbitals*. While this configuration is at a higher energy for an isolated C-atom, the overall energy is much lower (more stable) for CH_4, with four bonds, than for CH_2 with only two bonds. In valence bond theory, bonds are composed from an electron in a hybrid atomic orbital being shared with an electron in another atomic orbital.

Carbon and Hydrogen.

1.1 (Partner 1 should run the program and Partner 2 should record the information.) Activate **C** on the atom toolbar and draw methane in **ChemSk**. As you can see the program adds the

---

[3]For each atom the p$_x$, p$_y$, and p$_z$ orbitals are all equivalent in energy. The x, y, z-designations emphasize that each orbital sequesters electrons in a different region of space.

[4]A good statement of the Pauli Principle is that there is a zero probability that two electrons with the same spin can occupy the same region of space. This is also consistent with other statements, including that no electrons in the same atom can have the same values for all four quantum numbers; unique combinations of n, *l*, and m$_l$ define orbitals in different regions of space.

hydrogens as you draw. **Copy to 3D** and optimize[5] (three times). Then visualize as **Sticks** and try different display modes—**Sticks, Balls and Sticks,** and end with **Wireframe** and **with Dots.** Use the left mouse button to roll the molecule around on the screen (**3D Rotation** should be active). <u>Measure and record two of the C–H bond lengths to the nearest 0.1 Å.</u>

1.2   Click **Angle,** and click on the atoms of any set of connected H–C–H atoms in that order. The atom changes color when the mouse is positioned correctly over each atom, and a small notifier at the bottom of the screen shows which atoms have been picked. <u>Measure and record two different H–C–H angles to the nearest degree.</u> The angles can be justified by considering *molecular orbitals* that combine the sp[3] hybrid orbitals of C with the s-orbitals of H. The bonded atoms are at the vertices of a tetrahedron. Roll the molecule around on the screen to get a good sense of what this means. As you can see, the optimizer is not perfect, but the angles are close to the ideal tetrahedral angle of 109.5°.

1.3   Hold the **Ctrl** key and **A** to select everything on the screen and then hit the **Delete** key. Create the molecule, ethane: $H_3C–CH_3$ by clicking (once) and dragging on the drawing area. **Copy to 3D** and optimize the structure *(three times)*. Use the left mouse button to roll the molecule around on the screen. <u>Draw a Lewis electron dot structure consistent with the structure on the screen.</u> Then measure and <u>record the C–C and two of the C–H bondlengths to the nearest 0.1 Å.</u>

1.4   Notice the tetrahedral geometry about each carbon atom. <u>Record two different H–C–H and two C–C–H angles to the nearest degree.</u>

1.5   On the **View** menu check **Show Multiple Bonds.** In ChemSk delete ethane, draw it again, and click the center bond to create ethylene, $H_2C=CH_2$. **Copy to 3D** and optimize (three times). <u>Draw a Lewis structure</u> and measure and <u>record the C–C and two C–H lengths to the nearest 0.1 Å. Record two different H–C–H and two C–C–H angles to the nearest degree.</u> This molecule is also called ethene.

1.6   Delete ethylene and draw acetylene, $HC\equiv CH$, optimize, <u>draw Lewis structures and repeat the measurements. Explain the trend in C–C bond lengths for the three molecules.</u> This molecule is also called ethyne.

2. CONFORMATIONS AND VAN DER WAALS SURFACES

2.1   (Partners switch roles.) Redraw $H_3C–CH_3$ from scratch. **Copy to 3D** and optimize <u>one time only.</u>[5] Rotate the molecule to look down the C–C bond. <u>Do the hydrogens at each end *eclipse* each other or are they *staggered*?</u>

2.2   Run the optimizer a second time.[5] View the new result in **3D.** <u>What has changed?</u> It is often a good idea to run the optimizer a few times for it to converge on a final structure.

2.3   Go back to ChemSk, and create $H_2C=CH_2$, and $HC\equiv CH$,[5] on the same drawing board that already contains $H_3C–CH_3$. Hit **Ctrl A, Copy to 3D,** optimize each molecule three times, and display as **Sticks** and **with Dots.** Rotate them around and compare them visually. Now click **Auto Rotate and Change Style.** The *van der Waals surface* of the molecules is displayed **with Dots** and **Spacefill.** This surface represents the outer positions for the valence

[5]In this module, always answer **No** to **Remove Hydrogens?** when running the optimizer.

electrons. When the valence electrons of different molecules collide, they repel each other or react to make a new molecule. Interactions between the van der Waals surfaces of different parts of a molecule also contribute to the three-dimensional folding of proteins, nucleic acids, and other molecules.

2.4 In **ChemSk,** hit **Ctrl A** to select all atoms, then **File → Save As** with filename **2carbs,** then delete all atoms from the screen.

3. MOLECULAR SHAPES

The valence shell electron pair repulsion (VSEPR) theory helps to explain the shapes of molecules, particularly when d-atomic orbitals are part of a hybrid or when lone pairs of electrons occur. In the VSEPR theory electron pairs occupy regions of space as far as possible from each other. This minimizes the effect of repulsion between them.

3.1 (Partners switch roles.) Draw H_2O, visualize as **Sticks** in **3D** and optimize[5]. Measure and record the H–O–H bond angle. How does that compare with the value given in your textbook? Draw the Lewis structure for H_2O. Now turn on **with Dots.** What occupies the large space enclosed by the van der Waals surface that is away from the bonds?

3.2 Leave the same display mode, then delete H_2O in **ChemSk.** Draw NH_3, **Copy to 3D,** optimize[5] and visualize. Describe the shape of the molecule. Measure and record the H–N–H bond angles. How do they compare with the value given in your textbook? Draw the Lewis structure for NH_3. What occupies the large space enclosed by the van der Waals surface that is away from the bonds?

3.3 Copy to **ChemSk,** click on **H,** then draw another H–N bond. **Copy to 3D,** optimize[5] the structure and visualize it. What is the shape of the ammonium ion? Are all of the bonds equivalent? What happened to the large space in the van der Waals surface of ammonia?

3.4 (Partners switch roles.) Delete everything from the screen in **ChemSk,** draw and PCl_5, **Copy to 3D** and optimize[5] then visualize as sticks. This arrangement of five *ligands* about a central atom is called a *trigonal bipyramid.* Roll it around on the screen to be sure you can distinguish that there are two different classes of bond angles. The four atoms in the "trigonal" part of the molecule all lie in the same plane. Measure and record several bond angles in this planar part. Measure and record the other angles, as well.

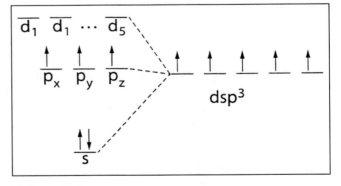

FIGURE 9-3

The ground state configuration of phosphorus is shown on the left. A d-orbital can mix with s- and three p-orbitals to make dsp^3 hybrid orbitals. Energy increases along the vertical axis.

3.5 Delete the previous molecule, draw SF_6, **Copy to 3D** and optimize[5] SF_6, then visualize as sticks. In this molecule the six ligands surround the central atom in an *octahedral* arrangement. Are all the bonds equivalent? What are the bond angles?

The arrangements of atoms in PCl_5, SF_6, and similar molecules cannot be explained by hybridization of s- and p-orbitals. The valence electrons of phosphorus and sulfur both have principal quantum number, n = 3, so that means that the second quantum number, l, can equal 0 (s), 1 (p), or 2 (d-orbitals). Thus, we should consider hybrids that involve the 3s, 3p, and 3d orbitals. Figure 9-3 illustrates that the d-orbitals are not occupied in phosphorus atoms in the ground state. However, five bonds can be explained if the phosphorus has the five dsp^3-orbitals illustrated at the right of the figure. A similar explanation applies to SF_6. Sulfur has one more valence electron than phosphorus, so the p_x-orbital in Figure 9-3 contains a ↑↓ pair of electrons. Then a six-lobed d^2sp^3-hybrid on sulfur can accommodate six ligands.

Lone pairs of electrons can occupy hybrid orbitals as well as bonding pairs. When you displayed ammonia, you noted that the van der Waals surface enclosed a large space that was not associated with bonding electrons. This extra *electron density* comes from a *lone pair* of electrons. These electrons are associated with only one positively charged nucleus, so lone pairs take up more space than they would if held in a bond. They tend to repel other lone pairs and bonding pairs, and compress the other bond angles. The parameters involved in optimizing structures in ChemSketch have not been fine-tuned to reproduce these effects. However, in real NH_3 molecules the H–N–H bond angles are smaller (107°) than the ideal tetrahedral angle (109.5°). Water is an example where there are lone pairs occupying two lobes of sp^3-hybrids centered on the oxygen. These two lone pairs compress the H–O–H bond angle to 104.5°.

4. MOLECULAR ORBITALS

The most comprehensive theory for chemical bonding is based on *molecular orbital theory*. Many of the details will already be familiar from the previous discussion. The tetrahedral arrangement in methane places the bonding electrons as far apart as possible in three-dimensional space. *A sigma bonding molecular orbital* (σ-MO) results from combining one of the four sp^3-hybrids from the carbon with the s-orbital of a hydrogen atom. In methane there are four equivalent bonding MO's, each occupied by two electrons. This is illustrated for one of the bonds in Figure 9-4. So far this is similar to the valence-bond theory. However, in addition to the sigma-bonding MO's there are sigma-antibonding MO's (σ*) that are not occupied in methane's *ground state* (the normal electronic state of the molecule in the absence of extreme heat or electromagnetic radiation). Electrons in bonding MO's favor the region between the two nuclei; the attraction of the nuclei for the bonding electrons holds the nuclei together. Electrons in antibonding MO's spend most of the time outside the internuclear region.[6]

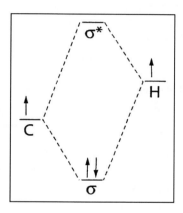

FIGURE 9-4
Two electrons occupy a sigma bonding molecular orbital in a C–H bond. C and H each contribute one electron to the bond. The bonded atoms have the electrons paired in the σ-MO. Energy increases along the vertical axis.

[6]An outcome of the quantum mechanical description of molecules is that the number of molecular orbitals must be the same as the number of atomic orbitals in the parent atoms.

4.1 In **ChemSk,** delete the screen contents and open the file, **2carbs.sk2,** view in **3D.** Use it as you read through the following description of bonding in ethane, ethylene and acetylene.

The same sigma bonding MO description discussed for methane applies in ethane. A σ bond between the carbons results from combining an sp^3-hybrid from each of the carbons. The other lobes on each carbon combine with the s-orbitals of hydrogen atoms to make additional σ bonds. Thus, all of the bond angles you measured for ethane are about 109°.

In ethylene the s- and two p-orbitals combine to make a three-lobed *sp^2-hybrid*, where the three bonds are as far apart as is possible in a plane (120°). These are involved in C–C and C–H σ bonds. All the nuclei in this sigma framework lie in the same plane, which can be assigned as the xy-plane in a Cartesian axis system. Roll the molecules on the screen to visualize this planar arrangement. As only three of the available hybrid orbitals in Fig. 9-1 are used for σ-bonds, one is left over for each carbon. The leftovers are the p_z-orbitals, which sequester electrons so they do not overlap with the sigma framework. The p_z-orbitals on the two carbons combine to make a *pi* (π) bonding molecular orbital, occupied by two electrons.[7] The σ and π bonds together make a C=C double bond.

In acetylene only one of the p-orbitals combines with the s-orbital to make a two-lobed *sp-hybrid*, where the C–C and C–H sigma bonds are as far apart as is possible in a line (180°). If this line is taken to define the x-axis, then the leftovers are the p_y- and the p_z-orbitals, which have their electron density as far away as possible from each other and the sigma framework. The p_y-orbitals from each carbon combine to make a π_y bonding MO, and the p_z-orbitals combine to make a π_z bonding MO.[7] The σ and two π bonds comprise a C≡C triple bond.

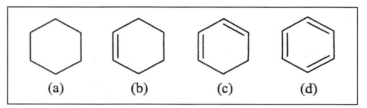

FIGURE 9-5
Carbon atom framework for (a) cyclohexane, (b) cyclohexene, (c) cyclohexadiene, (d) benzene.

5. DELOCALIZED BONDS

The electrons in MO's occupy the region between the bonded nuclei, but they also spend some of their time outside this region. This can have important effects, especially for π electrons. In the next set of exercises you will examine *conjugated* systems, which alternate double and single bonds, and *aromatic* rings where conjugation occurs completely throughout a cyclic molecule.

5.1 (Partners switch roles.) Delete the previous molecules in **ChemSk,** click **Templates,** and **Table of Radicals.** In the category called **Cycles,** move the mouse over the entries, click the one called **cyclohexane,** then click <u>once</u> in the drawing area. **Copy to 3D** and note that this looks like a flat hexagon. Optimize[5] the structure three times, visualize as **Wireframe** and **Resize** to make it as large as possible on the screen. Roll the molecule around on the screen and notice that it has a puckered, not flat, structure. This is because each of the carbons is sp^3-hybridized. Measure and <u>record two of the C–C bond lengths and C–C–H bond angles. Are they similar to the single bonded carbons in ethane?</u>

[7]As with σ- and σ^*-MO's, both π-bonding and π^*-antibonding MO's are created.

5.2 You will now make a molecule that looks like the one in Figure 9-5b. Rotate the molecule so that you are looking down on the ring and it looks like a hexagon. **Copy to ChemSk** and click on **Select/Move,** pick one of the H atoms and delete, then move to an adjacent carbon and delete one H there. Click **C,** then choose the bond between the carbons that have only one attached H, and make a double bond. This molecule is called *cyclohexene.* Optimize[5] and view in **3D.** Set to **Wireframe** and on the **View** menu be sure that **Show Multiple Bonds** is checked. Note that the puckering has changed and that the double bonded carbons and their attached H-atoms are all in the same plane (as in ethylene). Measure and <u>record the C=C bond length and the bond angles about the sp^2-hybridized carbons.</u> <u>Are they similar to those in ethylene?</u> Measure and <u>record the C–C bond length and two of the C–C–H bond angles on the opposite side of the ring from the double bond.</u> <u>Are they similar to the single bonded carbons in cyclohexane?</u>

5.3 *Conjugated double bonds.* In this part you will create another double bond as in 5.2. However, be sure that you skip one bond so the carbon framework is now like that in Figure 9-5c. Optimize[5] and view in **3D.** Roll the molecule around and note that the puckering has changed again; the double bonded carbons and their attached H-atoms are all nearly in the same plane. Measure and <u>record the C=C bond lengths and the bond angles about the sp^2-hybridized carbons.</u> <u>Are they similar to those in ethylene?</u> Measure and <u>record the C–C bond length farthest from the double bonds.</u> <u>Is it similar to those in cyclohexane?</u>

<u>Now measure the single bond between the two double bonds.</u> It is shorter than a typical C–C bond in cyclohexane, yet it is not a double bond. The bond is shorter due to *conjugation.* The effects of the π-bonds extend beyond the immediate region of the double bond. This is a good example of a molecular orbital being *delocalized*—belonging to the molecule as opposed to belonging just to individual bonds.

5.4 *Aromatic systems.* An important class of molecules extends conjugation a step further. In **ChemSk** add one more double bond so the ring now looks like Figure 9-5d. Optimize[5] and view in **3D.** Roll the molecule around and note that the puckering has changed again—the molecule is now completely flat—all of the carbons are sp^2-hybridized. Measure and <u>record the C=C and C–C bond lengths and a few of the C–C–C and C–C–H bond angles.</u>

How can the lengths all be the same? The answer is that there is an endless cycle of conjugation, where the π-electrons are delocalized in a molecular orbital that belongs to the whole molecule. Every carbon is equivalent to every other carbon and every carbon-carbon bond is equivalent to every other one. The bond lengths in aromatic systems are shorter than single bonds, but longer than double bonds.

This structure can also be explained in terms of *resonance,* where the cycle of double bonds in Figure 9-5d is moved clockwise by one bond, and the set of bonds is imagined to jump back and forth between the two forms. The concept of resonance has some value in pencil and paper analysis of molecular structure, but is not necessary if one focuses on delocalized molecular orbitals.

REPORT

Turn in your report at the end of the lab period today. Have your instructor initial the report before you submit it. Your instructor will tell you where and when you can pick up your graded report.

Periodic Table of the Elements

1	2	3	4	5	6	7	8	9	10	11	12	13	14	15	16	17	18
1 **H** 1.01																	2 **He** 4.003
3 **Li** 6.94	4 **Be** 9.01											5 **B** 10.81	6 **C** 12.01	7 **N** 14.01	8 **O** 16.00	9 **F** 19.00	10 **Ne** 20.18
11 **Na** 22.99	12 **Mg** 24.30											13 **Al** 26.98	14 **Si** 28.09	15 **P** 30.97	16 **S** 32.07	17 **Cl** 35.45	18 **Ar** 39.95
19 **K** 39.10	20 **Ca** 40.08	21 **Sc** 44.96	22 **Ti** 47.88	23 **V** 50.94	24 **Cr** 52.00	25 **Mn** 54.95	26 **Fe** 55.85	27 **Co** 58.93	28 **Ni** 58.69	29 **Cu** 63.55	30 **Zn** 65.39	31 **Ga** 69.72	32 **Ge** 72.61	33 **As** 74.92	34 **Se** 78.96	35 **Br** 79.90	36 **Kr** 83.80
37 **Rb** 85.47	38 **Sr** 87.62	39 **Y** 88.91	40 **Zr** 91.22	41 **Nb** 92.91	42 **Mo** 95.94	43 **Tc** 97.91	44 **Ru** 101.1	45 **Rh** 102.9	46 **Pd** 106.4	47 **Ag** 107.9	48 **Cd** 112.4	49 **In** 114.8	50 **Sn** 118.7	51 **Sb** 121.8	52 **Te** 127.6	53 **I** 126.9	54 **Xe** 131.3
55 **Cs** 132.9	56 **Ba** 137.3	71 ***Lu** 175.0	72 **Hf** 178.5	73 **Ta** 180.9	74 **W** 183.8	75 **Re** 186.2	76 **Os** 190.2	77 **Ir** 192.2	78 **Pt** 195.1	79 **Au** 197.0	80 **Hg** 200.6	81 **Tl** 204.4	82 **Pb** 207.2	83 **Bi** 209.0	84 **Po** 209.0	85 **At** 210.0	86 **Rn** 222.0
87 **Fr** 223.0	88 **Ra** 226.0	103 **§Lr** 262.1	104 **Rf** 261.1	105 **Ha** 262.1	106 **Sg** 263.1	107 **Ns** 262.1	108 **Hs** 265.1	109 **Mt** 266.1									

* Lanthanides

57 ***La** 138.9	58 **Ce** 140.1	59 **Pr** 140.9	60 **Nd** 144.2	61 **Pm** 144.9	62 **Sm** 150.4	63 **Eu** 152.0	64 **Gd** 157.2	65 **Tb** 158.9	66 **Dy** 162.5	67 **Ho** 164.9	68 **Er** 167.3	69 **Tm** 168.9	70 **Yb** 173.0

§Actinides

89 **§Ac** 227.0	90 **Th** 232.0	91 **Pa** 231.0	92 **U** 238.0	93 **Np** 237.0	94 **Pu** 244.1	95 **Am** 243.1	96 **Cm** 247.1	97 **Bk** 247.1	98 **Cf** 251.1	99 **Es** 252.1	100 **Fm** 257.1	101 **Md** 258.1	102 **No** 259.1

Notes

Notes

Laboratory 10

Chemistry of Recycling Aluminum

INTRODUCTION

Aluminum is a ubiquitous material in today's world and is the third most abundant element on the planet. It is found in cars, bicycles, airplanes, heatsinks, and virtually anywhere a lightweight, non-corroding metal is required. Maybe most common is the aluminum can, used for soda. In this experiment you will be recycling aluminum scrap in a very unusual way and you will produce two very useful products: hydrogen gas (H_2) and very pure hydrated potassium aluminum sulfate, $KAl(SO_4)_2 \cdot 12H_2O$, or alum. Hydrogen gas has great potential use as a fuel for combustion engines and fuel cells. Alum is a widely used chemical in industry, playing an important role in the production of many products used in the home and industry.

Alums have a wide range of commercial uses. Sodium alum is combined with sodium bicarbonate to make baking powder. The slight acidity of Al^{3+} in moist dough causes an acid-base reaction producing CO_2 gas, thus causing the rising of dough in baked goods with light, airy textures. Alums are also used in anti-perspirants/deodorants to kill odor-producing bacteria and plug sweat glands. Over 70% of all alums produced annually are used in paper factories to help cellulose molecules stick more tightly to one another to make stronger paper, a process known as sizing. Other uses include soaps, greases, fire extinguisher compounds, textiles, leather, synthetic rubber, drugs, cosmetics, cement, plastics, and pickles.

It is important to note here that the particular process for converting aluminum into alum being used in lab would produce very expensive alum! Today, alum is produced very cheaply using clay as the raw material. Consequently, the procedure used in this experiment is not used as an industrial method for recycling aluminum, but is still synthetically valid.

I. THE CHEMISTRY OF THE EXPERIMENT

The common aluminum can has a thin polymer coating on the inside of the can to protect it from corrosive liquids in the can. The outside of the can has layer of paint that also protects it from corrosion. The outer paint layer must be removed before beginning the recycling process. You

Reprinted by permission of Dr. Neal Abrams.

will accomplish this using a sanding sponge. The cleaned piece of aluminum is then dissolved in potassium hydroxide, KOH, a strong base. The reaction proceeds according to the following balanced reaction:

$$2 Al_{(s)} + 2 KOH_{(aq)} + 6 H_2O_{(l)} \rightarrow 2 KAl(OH)_{4(aq)} + 3 H_{2(g)}$$

This is an oxidation-reduction reaction whereby aluminum is oxidized from Al^0 to Al^{3+} and the hydrogen atoms in H_2O and KOH are reduced to H_2 gas. The clear and colorless $Al(OH)_4^-$ complex that is formed is known as an aluminate and explains why alkaline solutions (like Drano) are never stored in aluminum containers. The can would slowly dissolve!

Once the $KAl(OH)_4$ is dissolved, sulfuric acid is added to complete a sequential two-step reaction:

$$2 KAl(OH)_{4(aq)} + H_2SO_{4(aq)} \rightarrow 2 Al(OH)_{3(s)} + 2 H_2O_{(l)} + K_2SO_{4(aq)} \qquad \textbf{(Step 1)}$$

The reaction above is an acid-base reaction in which the H^+ ions from the sulfuric acid neutralize the base $Al(OH)_4^-$ to give a thick, white, gelatinous precipitate of aluminum hydroxide, $Al(OH)_3$. As more (excess) sulfuric acid is added, the precipitate of $Al(OH)_3$ dissolves to form soluble aluminum sulfate:

$$2 Al(OH)_{3(s)} + 3 H_2SO_{4(aq)} \rightarrow Al_2(SO_4)_{3(aq)} + 6 H_2O_{(l)} \qquad \textbf{(Step 2)}$$

The overall net ionic equation for Step 2 is:

$$Al(OH)_{3(s)} + 3 H^+_{(aq)} \rightarrow Al^{3+}_{(aq)} + 3 H_2O_{(l)}$$

This yields aluminum ions, Al^{3+}, in solution. The solution at this point contains Al^{3+} ions, K^+ ions (from potassium hydroxide), and SO_4^{2-} ions (from sulfuric acid). On cooling, crystals of **hydrated** potassium aluminum sulfate, $KAl(SO_4)_2 \bullet 12\ H_2O$ (or alum) are very slowly precipitated out of solution. In the experiment the rate of crystallization process can be increased by providing a small "seed crystal" of alum for the newly forming crystals to grow on. Cooling is needed because alum crystals are soluble in water at room temperature. The complete equation is:

$$Al_2(SO_4)_{3(aq)} + K_2SO_{4(aq)} + 24 H_2O_{(l)} \rightarrow 2 KAl(SO_4)_2 \bullet 12 H_2O_{(s)}$$

The net ionic equation is:

$$K^+_{(aq)} + Al^{3+}_{(aq)} + 2 SO_4^{2-}_{(aq)} + 12 H_2O_{(l)} \rightarrow KAl(SO_4)_2 \bullet 12 H_2O_{(s)}$$

Here it is important to discuss the definition of a hydrate. As indicated above, alum has the formula $KAl(SO_4)_2 \bullet 12\ H_2O$, meaning it is hydrated with twelve molecules of water. It is possible for the alum molecule (and many others) to hold on to water even though the water molecule is not chemically bound to the structure. The twelve water molecules on alum define it as being hydrated and these water molecules must be included when calculating the total molecular mass of the compound. In many cases these "waters of hydration" can be driven off of a compound by heating to temperature $> 100°C$.

Finally, the crystals of alum are removed from the solution by vacuum filtration and washed with an alcohol/water mixture. This wash liquid removes any contamination from the crystals but does not dissolve them. It also helps to dry the crystals quickly, because alcohol is more volatile than water.

The objectives of this experiment are:
- To prepare a sample of alum from aluminum scrap
- To perform stoichiometric calculations that determine the percent yield of product
- Determine costs of recycling aluminum
- Grow crystals of potassium alum, $KAl(SO_4)_2 \cdot 12\,H_2O$

REFERENCE

Silberberg, 2nd ed., Chapter 3, *Stoichiometry of Formulas and Equations*

2. MATERIALS

Aluminum can (bring your own)
Potassium hydroxide pellets
9M sulfuric acid, H_2SO_4
Hotplate
Filter paper
Filter flask
50:50 Ethanol:water solution
Supplies from lab drawer

SAFETY

You will be using solutions with high concentrations of sulfuric acid and potassium hydroxide, both of which are highly damaging to skin and eyes. Be careful when handling them. If you spill any on yourself, wash it off with lots of water. Neutralize any spills on the counter with baking soda.

When aluminum dissolves in potassium hydroxide solution, hydrogen gas is produced. Make sure that no flames are present. This step needs to be performed in a fume hood.

Boiling concentrated solutions often results in bumping and spurting of the hot liquid. Protect yourself by heating inside a hood with the **sash lowered**. Wear goggles and gloves.

---

3. EXPERIMENTAL PROCEDURE

Preparing the Aluminum

3.1 Using the can you brought to lab, pierce the can with the point of a pair of scissors and cut around so that the sides of the can are cut out. Deposit the waste aluminum scraps left over in the box provided.

3.2 Lay the rectangular piece of aluminum (the sides of the can) on the bench and scour both sides with the scrubber provided. Make sure that a 5cm x 5cm area is clean on both sides.

3.3 Wipe the metal clean with a paper towel and cut out a clean piece that is about 5 cm x 5cm.

3.4 Obtain the mass of the piece using the balance. Weigh the piece accurately, and record the mass in your notebook.

3.5 Cut the weighed piece into smaller pieces, and place them a clean 250 mL beaker. The smaller the pieces, the faster the reaction, but do not lose any metal bits during the transfer.

Dissolving the aluminum forming a soluble aluminate

3.6 Prepare 25 mL of a ~3M KOH solution using the information from the prelab question. This solution does not need to be exact and therefore does not require the use of a volumetric flask. A graduated cylinder should suffice for measuring the 25 mL of water and the solution can be prepared in a beaker. Add this solution to the aluminum IN THE HOOD!

3.7 Place the beaker on the hotplate over low heat. It is not necessary to boil the solution. The aluminum will take about 20 minutes to dissolve.

3.8 While it is dissolving, set up an apparatus for gravity filtration.

3.9 When the aluminum has dissolved (as evidenced by the lack of bubbles of H_2 gas given off), gravity filter the solution. Only fill the funnel to within 1/2" of the top of the paper. Use a glass rod to decant the solution into the paper, leaving behind any bits of plastic coating and/or paint. The solution in the Erlenmeyer flask should be both clear and colorless at this point.

Forming insoluble Al(OH)₃, neutralizing OH⁻, and forming soluble Al₂(SO₄)₃

3.10 Allow the flask to cool. When the solution is reasonably cool, add 20 mL of 9 M H_2SO_4 slowly and with care. It is important that you and your lab partner swirl the flask as you add the acid. The solution will get quite warm. If there are any white flecks left in the solution after the addition of the H_2SO_4, place the flask on a hotplate set to low and warm it with swirling until all of the solid material has dissolved.

  • Record the changes that take place during the addition of the H_2SO_4.

   — Becomes solid / white. Foggy

Forming alum crystals

3.11 Make an ice bath for the flask by putting ice and water into a beaker. Allow the flask to cool on the bench for 5 minutes, then place it in the ice bath and allow it to cool for an additional 5 minutes. Record any observations over time.

3.12 Solid alum crystals should begin to form. If they have not, scratch the inside walls of the flask with a glass stirring rod. This provides nucleation sites where crystallization can begin, followed by crystal formation throughout the liquid. Swirl the flask when you notice the onset of crystal formation and allow it to cool in the ice bath for another 10 minutes.

3.13 While the solution is cooling, pour 25 mL of 50:50 ethanol:water mixture into an Erlenmeyer flask and place it in the ice bath to cool.

3.14 Set up the vacuum filtration apparatus (using a Büchner funnel and filter paper). Pour some deionized water onto the filter paper.

3.15 Remove the flask containing the alum crystals from the ice bath, swirl so that all the crystals are suspended, and pour quickly into the Büchner funnel. Keep swirling and pouring until all the solution and crystals are transferred to the funnel. The vacuum should be kept going throughout this process.

3.16 Pour about 10 mL of the cold ethanol/water mixture into the flask with the remaining alum. Swirl the flask and pour the mixture into the funnel to transfer any remaining crystals.

3.17 Wash the alum crystals, now on the filter, with the cooled alcohol/water mixture using the following procedure:

- Disconnect the vacuum hose from the vacuum port. Pour about 5-10 mL of ethanol/water onto the crystals and gently swirl. Reconnect the aspirator hose, and suck the crystals dry. After pouring on the last portion of alcohol/water, leave the vacuum on for at least 20 minutes.

3.18 Calculate the theoretical yield of alum while the crystals are drying and review work as described by the instructor.

3.19 Transfer your crystals to a weigh boat and accurately weigh your yield of crystals, recording the weight of product in your lab notebook. Be careful not to scrape the filter paper into your crystal yield! —

3.20 After weighing your alum sample (*make sure you have recorded the mass!*), give the sample to your TA and record the initial mass of the aluminum and final mass of alum on the sheet provided.

Mass$_i$ Mass$_f$

1.124g   9.529g

Cleanup

Pour any excess solutions into the designated waste container. Place solid materials in the designated waste container. Wash all glassware with soap and water and conclude with a DI water rinse.

4. REACTION ANALYSIS

The stoichiometry involved in the sequence of reactions yields the mole relationship between aluminum and alum. This is required to calculate the percentage yield for your product.

$2\ Al(s) + 2\ KOH + 6\ H_2O \rightarrow 2\ K[Al(OH)_4] + 3\ H_2$	(1)
$2\ K[Al(OH)_4] + H_2SO_4 \rightarrow 2\ Al(OH)_3(s) + 2\ H_2O + K_2SO_4$	(2)
$2\ Al(OH)_3 + 3\ H_2SO_4 \rightarrow Al_2(SO_4)_3 + 6\ H_2O$	(3)
$Al_2(SO_4)_3 + K_2SO_4 + 24\ H_2O \rightarrow 2\ KAl(SO_4)_2 \bullet 12H_2O$	(4)
$2\ Al(s) + 2\ KOH + 4\ H_2SO_4 + 22\ H_2O \rightarrow 2\ KAl(SO_4)_2 \bullet 12H_2O + 3\ H_2$	(5)
The **overall reaction** for the synthesis of alum, equation (5), is obtained by adding reactions (1-4) and canceling species that appear on both sides of the equation. The overall reaction stoichiometry (5) signifies that 2 moles of aluminum will produce 2 moles of alum.	

Notes

For measuring out H_2SO_4, use small beakers labeled H_2SO_4 + transfer from container to graduated cylinder.

Between step _3.10 - 3.11_ boil off some H_2O.

Making 3M KOH.

so $\dfrac{x\, mols}{.025\, L} = 3$

K al (OH)$_2$ $x\, mols = 3 \times .025L$

$0.075\, mols \times \left(\dfrac{56g}{1\, mol}\right)$

$= 4.2g$

Notes

Laboratory 11

Preparation and Viscosity of Biodiesel from Vegetable Oil*

INTRODUCTION

The term "renewable energy" has become a ubiquitous part of the English vocabulary here at ESF as well as the entire nation. As fossil fuel prices are on the rise, society *and* science are searching for a comparable alternative. Some speculate that we have already reached our peak output for fossil fuels and quantities will continue to decline. This leaves us, as scientists, to develop new sources of fuel, preferably based upon renewable resources. The development of biofuels has been at the forefront of scientific research, including ESF's active research program aimed at converting woody biomass to ethanol. Among the many possible alternative fuel choices, biodiesel has emerged as a popular choice because of the low cost of production and ease of integration into today's diesel engines. Biodiesel also has the benefit of not containing any nitrogen or sulfur compounds, so it does not contribute to acid rain or smog production. As of 2006, 23 states have passed legislation regarding the use or introduction of biodiesel as an alternative fuel. Biodiesel has also been recognized by the EPA as a safe and legal fuel for use in combustion engines.

Biodiesel is produced by the catalytic transesterification of regular vegetable oils, such as canola or soybean oil. Although biodiesel looks to be a promising alternative to fossil fuels, it does have some drawbacks. One of these is the high viscosity and solidification of biodiesel at low temperatures (think Crisco). How do you think this would influence biodiesel use in Central New York? One common way of alleviating this issue is to combine biodiesel with petroleum-based diesel in a blend known as B20 (20% biodiesel, 80% diesel). Why do you suppose combining diesel with biodiesel prevents solidification? Is this a good solution?

In this laboratory exercise, you will work toward synthesizing biodiesel, an alternative fuel made from vegetable oils. Biodiesel is the methyl ester made from the transesterification of the

* Adapted from: Nathan R. Clarke, John Patrick Casey, Earlene D. Brown, Ezenwa Oneyma and Kelley J. Donaghy, 2006, *J. Chem. Ed.*, 83(2), 257-259.

117

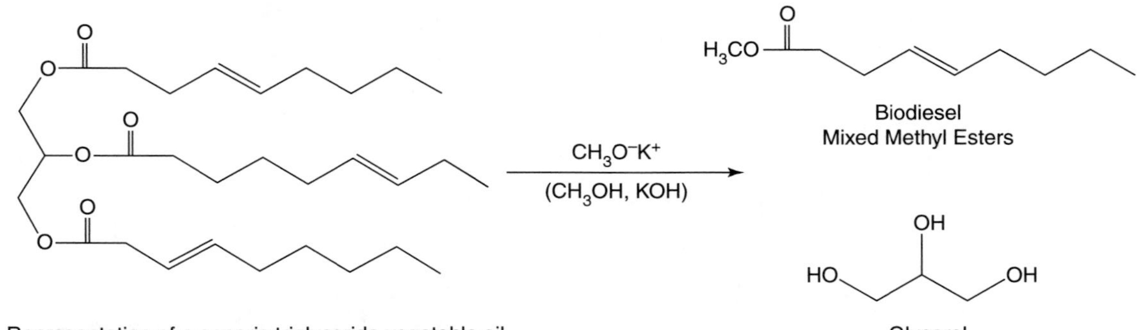

Representation of a generic triglyceride vegetable oil

Biodiesel Mixed Methyl Esters

Glycerol

Scheme 1.

Transesterification of a Generic Triglyceride. Notes: For real oils such as soybean or canola oil, the chains of carbon atoms shown are actually 16-18 atoms long. The catalyst for this reaction is potassium hydroxide, which makes potassium methoxide ($CH_3O^-K^+$) in methanol (CH_3OH).

triglycerides present in vegetable oils as shown in **Scheme 1**. In addition to experimenting with organic synthesis, you will use some simple viscosity measurements to gain an understanding of the intermolecular properties of biodiesel to appreciate how it flows at low temperature.

1. GOALS

In this experiment you will:

1.1 Prepare a sample of biodiesel fuel from soybean oil by transesterification with methanol.
1.2 Measure the relative viscosity of biodiesel.
1.3 Report on the viability and potential problems you have found for the widespread use of biodiesel fuel using this synthetic method.

2. REAGENTS

Methanol
Potassium hydroxide pellets, KOH
Vegetable oil

3. EQUIPMENT/MATERIALS

Bottle with cap
Hot plate
Graduated cylinder, 100 mL
Glass stirring rod
Pasteur pipettes
Stopwatch
150 mL Beaker

SAFETY PRECAUTIONS

- Safety goggles and gloves must be worn when working with chemicals and materials.
- **CAUTION**: Avoid direct inhalation of the potassium methoxide.
- If a spill occurs, immediately contact a TA for appropriate clean up procedures.
- Potassium hydroxide is caustic; methanol, biodiesel, B20 are flammable; all should be handled with care.
- The addition of the sodium hydroxide / methanol to the warm oil should be done slowly and with care.

---

4. EXPERIMENTAL DETAILS

Part 1 - Biodiesel Synthesis

Step 1. You will use approximately 100 mL of canola oil, which is a renewable agricultural product (approx. molar mass = 880 g/mol; density = 0.917 g/mL). To assure complete trans-esterification, 0.59 g of potassium hydroxide is required per 100 milliliters of oil. The potassium hydroxide is dissolved in methanol to yield potassium methoxide. Use the answers from the prelab to calculate to prepare the 0.4 M potassium methoxide solution by combining the KOH and methanol in a capped plastic bottle (*assume the methanol is the only contribution to volume*). Cap the solution and **swirl** (*do not shake*) until the KOH is dissolved. *Caution: This solution is caustic and flammable. Be very careful when handling it.* Once no solid particles remain, you have potassium methoxide ready to be used.

Step 2. Prepare a water bath and heat the water to 50°C.

Step 3. Once the bath temperature is stable (within 2-3°C), add the canola oil (100 mL) to a 150 mL beaker. Place the beaker in the water bath and heat the canola oil to 50°C and *slowly* add the calculated amount of potassium methoxide solution (use a glass stirring rod to stir the solution).

Step 4. Raise the temperature of the water bath to 65-70°C, maintained a constant stirring rate. Place a watch glass on the beaker to minimize evaporation.

Step 5. Allow the reaction to proceed with *constant* stirring and constant heat for 45 minutes. You will need to constantly monitor the temperature of the bath. You may move on to Part II as long as someone monitors the temperature.

Step 6. After 45 minutes, remove the reaction beaker from the bath and allow it to cool without stirring. **BE CAREFUL! THIS IS HOT!** The solution will separate into two phases, glycerol and biodiesel. Which phase is which knowing that glycerol is denser than biodiesel?

Step 7. Once the temperature cools to approximately 40°C, decant the biodiesel (not the glycerol) into the washed 100 mL graduated cylinder originally used to measure out the oil. Again, do not transfer the glycerol. Record the volume of the biodiesel.

Step 8. See an instructor regarding testing the flammability of your biodiesel fuel.

Step 9. Pour the biodiesel fuel into the large collection cylinder. Place the residual glycerol in the specified waste container.

Part II - Viscosity of Biodiesel

Determine the relative viscosities of the product, B20 blend, and the starting oil. The relative viscosities can readily be determined by timing the passage through a calibrated, 6" Pasteur pipette using biodiesel, followed by the original oil and B20 blend.

Step 1. The "calibration" consists merely of putting a mark on the body of the pipette with a marker. A second mark should be made on the narrow stem 2 cm up from the tip.

Step 2. With a finger held against the tip, fill the pipette with previously prepared biodiesel slightly above the calibration.

Step 3. Remove your finger and note the start time when the meniscus reaches the top mark. As the last of the biodiesel reaches the lower mark, the time is again noted. The biodiesel can drain into a waste beaker. Perform two additional trials for a total of three measurements. Record the data for theses trial is recorded in Table 1 of the Lab Report.

Step 4. Repeat the process using the vegetable oil starting material, obtaining three trials. Record the data for theses trial is recorded in Table 1 of the Lab Report.

Step 5. Repeat the process using a sample of B20, a blend of 20% biodiesel fuel and 80% regular petroleum-based diesel fuel. Again, obtain three trials of each of these. Record the data for theses trial is recorded in Table 1 of the Lab Report.

Step 6. The relative viscosity can be calculated as:

(Relative Viscosity) = (oil time)/(biodiesel time).

Be sure to use the same calibrated Pasteur pipette for all three measurements. Based on your qualitative observations on appearance, draw some conclusions about the viscosity of the various biodiesel samples compared to its properties at room temperature.

Clean-Up and Waste Disposal

- All synthesized oil should be poured into the large container for later use by the campus.
- Pour the fuels from the Part II, Viscometry, into the designated waste containers.
- All other waste (KOH, methanol) should be placed in the designated waste container.
- An especially thorough wipe down of your area is important in this lab. Even a small amount of base can burn the next person to come into contact with the surface. Use a sponge and a good amount of water. Rinse and wring out the sponge well after use.
- Rinse all glassware with isopropanol (IPA) and pour into the waste container. Follow up with washing using soap, water, and a DI rinse. Put beakers and test tubes in the inverted position in your drawer to dry.

REFERENCES

1. Deffeyes, Kenneth, S. Hubbert's Peak: The Impending World Oil Shortage. Princeton University Press, 2003.
2. The website of the National Biodiesel Board; http://www.biodiesel.org (accessed December 2007).

Notes

Waste 3 bottles

   glycerol (bottom layer)

Good biodiesel — show me!

   mess ups

0.4M potassium methoxide

$0.59g\ KOH \times \dfrac{1\ mol}{56.11\ g} = 0.0105\ mol\ KOH$

$0.4M = \dfrac{0.0105\ mol}{x\ L}$

$x = 0.0263\ L = \boxed{26.3\ ml}$

step 3 - stir every 5 minutes

✱ Skip step 6

Complete part II during step 5

✱ DO NOT OVERHEAT

CRUYO 11/15/12

Notes

Laboratory 12

Generating Hydrogen Gas

INTRODUCTION AND THEORY

The standard molar volume (V_{MOLAR}) of a gas is the volume occupied by <u>one mole</u> of the gas at standard conditions of temperature and pressure (STP).

> **Standard Temperature (T):** 0.00 °C, which is equal to 273.15 K.
> *Conversion:* K = °C + 273.15 K
>
> **Standard Pressure (P):** 760.0 mmHg (millimeters of mercury), which is equal to 1 atm.
>
> **Standard Molar Volume (V_{MOLAR}):** V_{MOLAR} of a gas at STP conditions is 22.4 liters per Mole (L mol^{-1}). In other words, one mole of gas at STP occupies a volume of 22.4 liters.

In this experiment, the molar volume of hydrogen gas at STP conditions will be determined through the reaction of magnesium metal with hydrochloric acid. The products formed are aqueous magnesium chloride and hydrogen gas. The balanced chemical reaction is as follows:

$$Mg_{(s)} + 2\ HCl_{(aq)} \longrightarrow MgCl_{2\ (aq)} + H_{2\ (g)}$$

Magnesium is in Group IIA of the periodic table and forms the Mg^{2+} cation in solution. Chlorine is in Group VIIA of the periodic table and forms the Cl^- anion in solution.

For Example:

A 0.0750 gram sample of magnesium metal react with hydrochloric acid to produce 77.5 mL of hydrogen gas. The gas is collected over water at 20.00 °C and an atmospheric pressure of 763.0 mmHg. Calculate the experimental molar volume of hydrogen gas at STP conditions.

Reprinted by permission of Dr. Arthur Stipanovic.

Follow the mathematical outline below to determine the molar volume (V_{MOLAR}) of hydrogen gas at STP conditions for the given example

1. **Calculate the temperature of the hydrogen gas (T_{H2}) in units of degrees Kelvin (K).**

   - In the example (*and in this experiment*), the hydrogen gas is collected over water. Therefore, the temperature of the hydrogen gas is assumed to be equal to the temperature of the water in the beaker.
   - The experimental temperature is measured in units of °C and must be converted into units of degrees K.

   $T_{H2} = °C + 273.15\ K$

   $T_{H2} = 20.00\ °C + 273.15\ K$

   $T_{H2} = 293.15\ K$

2. **Calculate the vapor pressure of the hydrogen gas (P_{H2}).**

   - In the example (*and in this experiment*), the hydrogen gas is collected over water. Therefore, both hydrogen gas ($H_{2(g)}$) and water vapor ($H_2O_{(g)}$) contribute to the total pressure. The total pressure (P_{TOTAL}) is equal to the atmospheric pressure (P_{ATM}) measured with a barometer.
   - In the example, the atmospheric pressure (P_{ATM}) is given as 763.0 mmHg. Table 12.1 lists the vapor pressure of water (P_{H2O}) at 20 °C as 17.5 mmHg.

Table 12.1 Vapor Pressure of Water at Various Temperatures

[T]	[P_{H2O}]	[T]	[P_{H2O}]	[T]	[P_{H2O}]
15 °C	12.8 mmHg	20 °C	17.5 mmHg	25 °C	23.8 mmHg
16 °C	13.6 mmHg	21 °C	18.6 mmHg	26 °C	25.2 mmHg
17 °C	14.5 mmHg	22 °C	19.8 mmHg	27 °C	26.7 mmHg
18 °C	15.5 mmHg	23 °C	21.1 mmHg	28 °C	28.3 mmHg
19 °C	16.5 mmHg	24 °C	22.4 mmHg	29 °C	30.0 mmHg

$P_{ATM} = P_{H2} + P_{H2O}$ ————rearrange the equation————> $P_{H2} = P_{ATM} - P_{H2O}$

$P_{H2} = P_{ATM} - P_{H2O}$

$P_{H2} = 763.0\ mmHg - 17.5\ mmHg$

$P_{H2} = 745.5\ mmHg$

3. **Calculate the volume of hydrogen gas at STP conditions (V_{STP}).**
   - In the example (*and in this experiment*), hydrogen gas ($H_{2\ (g)}$) is produced and collected as the chemical reaction proceeds. In the example, the volume of hydrogen gas produced (V_{H2}) is given as 77.5 mL.
   - The volume of hydrogen gas at STP conditions (V_{STP}) can be calculated using the **combined gas law**. In the equation on the following page, subscript "1" refers to the *experimental* conditions for hydrogen gas ($H_{2(g)}$) and subscript "2" refers to the *standard* conditions of temperature and pressure (STP).

$$\frac{P_1 V_1}{T_1} = \frac{P_2 V_2}{T_2}$$

More specifically

$$\frac{P_{H_2} V_{H_2}}{T_{H_2}} = \frac{P_{STP} V_{STP}}{T_{STP}}$$

Rearrange equation to solve for (V_{STP})

$$V_{STP} = \frac{P_{H_2} V_{H_2}}{T_{H_2}} \times \frac{T_{STP}}{P_{STP}}$$

Further re-arrangement

$$V_{STP} = V_{H_2} \times \frac{P_{H_2}}{P_{STP}} \times \frac{T_{STP}}{T_{H_2}}$$

$$V_{STP} = V_{H_2} \times \frac{P_{H_2}}{P_{STP}} \times \frac{T_{STP}}{T_{H_2}}$$

$$V_{STP} = 77.5 \, mL \times \frac{745.5 \, mmHg}{760.0 \, mmHg} \times \frac{273.15 \, K}{293.15 \, K}$$

$$V_{STP} = 70.8348...$$
$$V_{STP} = \underline{\underline{70.8 \, mL}}$$

4. **Calculate the number of moles of hydrogen gas (moles H$_2$) produced for the reaction.**
   - Refer to the balanced chemical reaction between magnesium metal and hydrochloric acid (previous page); magnesium metal and hydrogen gas are in a 1:1 mole ratio.
   - In the example (*and in this experiment*), the number of moles of hydrogen gas produced from the chemical reaction is based on the *initial amount of magnesium metal* that reacted. In the example, a 0.0750-gram sample of magnesium metal reacted.

$$0.0750 \, grams \, Mg \times \frac{1 \, mole \, Mg}{24.30 \, grams \, Mg} \times \frac{1 \, mole \, H_2}{1 \, mole \, Mg} = 0.003086... \longrightarrow 0.00309 \, moles \, H_2$$

5. **Calculate the molar volume of hydrogen gas (V$_{MOLAR}$).**
   - The ***molar volume*** of a gas is defined as the volume occupied by one mole of gas at STP conditions. It is a ratio of the volume of hydrogen gas (*in liters*) to the number of moles of hydrogen gas produced from the chemical reaction:

$$V_{molar} = \frac{Volume \, of \, H_2 \, gas \, at \, STP}{moles \, of \, H_2 \, gas \, produced} \times \frac{1 \, liter}{1000 \, mL}$$

Volume conversion

$$V_{molar} = \frac{70.8348... \, mL}{0.003086... \, moles} \times \frac{1 \, liter}{1000 \, mL}$$

$$V_{molar} = 22.9504...$$

$$V_{molar} = 23.0 \, L \, mol^{-1}$$

6. Calculate experimental error (%ERROR).

- The theoretical molar volume of hydrogen gas at STP conditions in 22.4 L mol[D1]. In the example, the molar volume of the hydrogen gas (V_{MOLAR}) was calculated to be 23.0 L mol^{-1}. Experimental error is calculated as follows:

$$\% \, error = \frac{|\, actual \, value - experimental \, value \,|}{actual \, value} \times 100$$

$$\% \, error = \frac{|\, 22.4 \, L \, mol^{-1} - 23.0 \, L \, mol^{-1} \,|}{22.4 \, L \, mol^{-1}} \times 100$$

$$\% \, error = 2.6785 \ldots \rightarrow 2.68 \, \%$$

EXPERIMENTAL PROCEDURE

SAFETY PRECAUTIONS:

- A safety goggles at all times in the lab. Gloves must be worn when working with hazardous chemicals.
- **CAUTION:** Hydrochloric acid is corrosive and can cause chemical burns. Handle the hydrochloric acid with care.

1. Obtain a piece of magnesium metal and record the mass to the nearest 0.001 gram.
2. Place approximately 250 mL of deionized water in a clean 400 mL beaker. Completely fill a clean 50 mL graduated cylinder to the rim with deionized water.

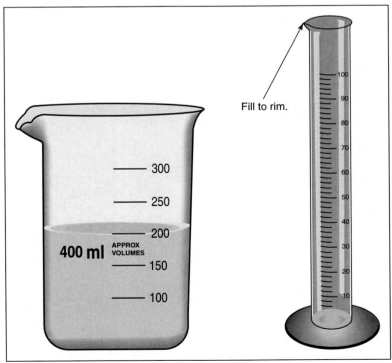

© Gjermund Alsos, 2012. Used under license from Shutterstock, Inc.

© Jan Kaliciak, 2012. Used under license from Shutterstock, Inc.

3. Wet a small piece of paper towel with deionized water. Place it over the rim of the graduated cylinder and remove any air bubbles. With a careful motion, quickly invert the graduated cylinder and place it in the beaker. Allow the piece of paper towel to float free and remove it.

   Note: It is imperative that there are no air bubbles in the cylinder once it is inverted; repeat this technique if there are any air bubbles.

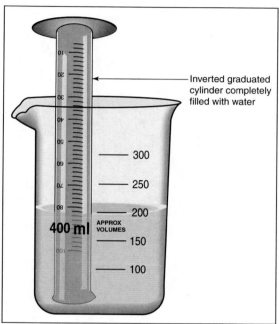

Beaker © Gjermund Alsos, 2012. Used under license from Shutterstock, Inc.

Graduated cylinder © Jan Kaliciak, 2012. Used under license from Shutterstock, Inc.

4. Slightly bend the piece of magnesium metal in half, but do not break it. Drop it into the water and adjust the graduated cylinder so that it is over the piece of metal.

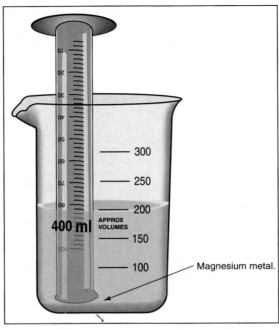

Beaker © Gjermund   Graduated cylinder ©
Alsos, 2012. Used under Jan Kaliciak, 2012.
license from   Used under license
Shutterstock, Inc.   from Shutterstock, Inc.

ATTENTION: WEAR GLOVES WHEN WORKING WITH HYDROCHLORIC ACID.

5. Obtain approximately 10 mL of 6M hydrochloric acid in a clean 50-mL beaker. Using a disposable plastic pipet, *slowly* add the acid to the water by the tip of the graduated cylinder at the bottom of the beaker.
   a. Do not allow air bubbles from the pipet to enter the graduated cylinder.
   b. Be careful not to disturb the positions of the metal and the cylinder as you add the acid.
6. Allow the magnesium metal to completely react. Record observations.
   a. Hold the cylinder in a vertical position over the metal to prevent loss of the formed gas.
   b. If the metal floats to the top of the cylinder during the reaction (since the bubbles sometimes stick to the side of the metal), then tap the side of the graduated cylinder so that the metal will sink back down into the acidic environment. Consult your TA is the floating persists.
7. Record the volume of hydrogen gas produced in Table 2 of the Lab Report to the nearest 0.5 mL. Be sure to record the temperature and atmospheric pressure in the room as well for later use in you calculations.
8. Repeat Steps 1-7 two more times (a total of three trials).
9. Complete the calculations in the Lab Report to obtain the average molar volume of the hydrogen gas and the % ERROR.

Notes

Avg. mass of $Mg = 0.04g$

$$0.04g = \frac{1 mol}{24.639g} = 0.0016 \, mol \; Mg$$

$Mol \; Mg = mol \; H_2 = 0.0016 \, mol \; H_2$

Next

Notes

0.045g

final volume 45mL

Avg. mass of Mg = 0.04g m_N $m_g = 24.305 \, g/mol$

$$0.04g * \frac{1 \, mol}{24.305g} = 0.0016 \, mol \, Mg$$

mol Mg = mol H_2 = 0.0016 mol H_2

$P_{H_2} = P_{atm} - P_{H_2O} \longrightarrow$ Table 12.1

$P_{H_2} = 757.4 - 19.8$ $P_{H_2} = 737.6 \, mm \, Hg$

avg. vol H_2 gas = 40 mL

$$\frac{P_{H_2} V_{H_2}}{T_{H_2}} = \frac{P_{STP} V_{STP}}{T_{STP}} \longrightarrow \frac{(737.6)(40 \, mL)}{295.4 \, K} = \frac{(760)(V_{STP})}{273.15 K}$$

$$V_{molar} = \frac{V_{STP}}{1 \, mol \, H_2} * \frac{1 \, L}{1000 \, mL}$$

$$V_{molar} = \frac{35.9 \, mL}{0.0016 \, mol} * \frac{1 \, L}{1000 \, mL} = 22.4 \, L/mol$$

% error

actual = 22.4 L/mol

$$\frac{22.4 - 22.4}{22.4} \times 100 = 0$$

CReyo
11/29/R

29.49 inches mercury
25.4 mm per 1 inch

↑
conversion for
atmospheric pressure